Troubleshooting Your Contracting Business To <u>Cause</u> Success

by
Charles Vander Kooi

Blue Willow, Inc.
Littleton, Colorado U.S.A.

Blue Willow, Inc.
P.O. Box 621227, Dept. 101
Littleton, Colorado 80162
(303) 932-1600
Web Page - http://www.bluewillow.com

Printed in the United States of America

Library of Congress Card Number: 98-70447

ISBN 1-889796-04-2

Preface and Acknowledgments

This book is the culmination of many years of the author's work within the construction industry. Literally thousands of people have shared ideas and helped solve problems during the course of hundreds of seminars and consultations.

Sincere appreciation must be expressed to Greg Ritscher who shared his many talents in the area of working with people. Greg co-authored the section "Charting the Right Course and Navigating With People." His insight and wisdom will help the reader prosper in one of life's greatest challenges, *working well with others*.

Special thanks to Howard Eckel, who has contributed his book "Growing and Staffing Your Business" to become the fourth section of this book. Howard has enjoyed tremendous success as an executive in the "Green Industry" and willingly shares his experience and insights with us to add to the "practical experience" nature of this book.

Additional thanks are due to the Ed and Belva Edwards, Tom and Mary Glover, and Mary Miller, who have invested many hours in the design, layout and production of this book.

Blue Willow, Inc. and Charles Vander Kooi have made a serious effort to ensure that information provided in this book is as accurate as possible. Thousands of person-hours have gone into the research, collection, verification, and editing necessary to complete this book; however, the publisher and author do not guarantee that all of the information included in this book is error free or exact. If you discover any errors, we would greatly appreciate it if you would report them to us. We also welcome any suggestions that you might have concerning future editions of any of our products.

Charles Vander Kooi Biography

Charles Vander Kooi has been involved in the construction industry for over thirty years, thirteen as an upper-management employee of companies and seventeen as a consultant. He has bid over a billion dollars in work over his career. As a private consultant, he has helped over 1,200 companies in their estimating/bidding systems and has lectured to over 70,000 contracting people nationally and internationally.

Constantly in demand, Mr. Vander Kooi speaks at an average of 100 Trade Shows and Conventions annually, teaching his philosophy across the U.S., Canada and Australia. His company consults with an average of 100 clients annually to assist in improving their performance.

He has authored eight books to the industry and his seminars are available on audio cassettes as well as videos.

Additional Products by Blue Willow. . .

Measure for Measure by Richard A. Young and Thomas J. Glover
ISBN 1-889796-00-X, 4"x6" soft cover, 864 pp, $14.95
The most complete conversion factor handbook ever written. Contains over 39,000 conversions for over 5,100 different units.

Seldovia, Alaska by Susan Woodward Springer
ISBN 1-889796-03-4, 8-1/2"x11," soft cover, 240 pp, $24.95
An historical portrait of life in the town of Seldovia, Alaska.

Pocket Partner by Dennis Evers, Mary Miller and Thomas J. Glover
ISBN 1-889796-02-6, 3.2" x 5.4", soft cover, 544 pp, $9.95
An amazing shirt pocket reference book for law enforcement personnel. Everything from bullet holes to hazmat to phones. *Scheduled for release in March 1998.*

Table of Contents

Table of Contents (cont.)

Table of Contents (cont.)

Table of Contents (cont.)

Table of Contents (cont.)

Table of Contents (cont.)

Food For Thought . . . (cont.)

Figures and Illustrations

Figures and Illustrations (cont.)

Growing and Staffing Your Business

Troubleshooting the Estimating and Bidding Process

Food For Thought

This Book Could Be The Most Important Investment You Ever Make In Your Business!

The Need for Troubleshooting Your Business to <u>Cause</u> Success

The major reason for business failures in this country is the lack of upper management paying attention to details. This book contains some of the major details you <u>must</u> pay attention to if you are to succeed.

No contractor of any type plans to fail in their business. They put far too much of themselves into their business to do that. Many, however, fail to plan to the degree that they insure success, or more predominantely, they fail to troubleshoot their original plan to overcome the inevitable weaknesses inherent in most original business plans.

Also, there are general principles applicable to business management and you must work with them to make them fit your particular company.

I have developed two terms which describe the underlying approaches most contractors take to their bidding process and, for that matter, to their entire company. I call them either a *"front door contractor"* or a *"back door contractor."* A *"back door contractor"* goes out and bids everything he can without any rhyme or reason. He approaches his company on a day to day, "face what comes my way" attitude. Whatever comes in the back door, on which he can be low bidder, dictates the kind and quality of work he can get, the people he has working for him and the kind of company he will be. That is a *"back door contractor."*

A *"front door contractor"* is the contractor who has a plan in mind. He knows his people and their abilities and he knows himself. He knows what he wants out of his company. He also knows what kind of work his company does best and what is most profitable. He sets up projections as to what his overhead will be and then establishes goals for his company and determines the amount of work required to recover that overhead. He then selectively goes out and gets that work. He brings in through the "front door" what he wants and needs. He is running his company rather than his company running him. That is a *"front door contractor."*

I read an alarming statistic the other day. It said that within the first three years of existence, 60% of the construction firms in this country become casualties. It went on to say that within ten years, 80% become casualties. Only 20% survive the long haul.

I have just put down a local newspaper which reported that a major construction company ($100 million per year) has just declared Chapter 11 bankruptcy. I am convinced, now more than ever before, that our industry needs more *"front door contractors."* I believe they will be the only ones who survive.

The principles and concepts I will be sharing with you in this book are things that can make you a *"front door contractor."* They are things I have seen and experienced while working both as an employee and as a consultant for several contractors.

A lot of what I will share with you in this book is very general in nature. Contractors of all different kinds and sizes will be reading it, so I cannot get down to the specific needs of your company. I will, however, share some basic principles which apply to all contracting companies, no matter what kind or size. It is your opportunity, <u>your responsibility,</u> to take these general principles and apply them to the particulars of your company.

Some of you may feel that your operation is too small to apply some of these principles. You may feel they are fine for a big company, but they are not necessary for yours. These principles are like the foundation of a building. If you don't build a solid foundation for your company when you are small, you will be trying to build the house of a bigger company on your weak foundation or you will be trying to rebuild that foundation while your house is being built. If you start right, you will finish right. No company is too small to do it right.

WARNING: Throughout this book I will be using examples which have either numbers or percentages in them. These numbers and percentages are made up for the purpose of examples and have nothing to do with national averages or other companies.

DO NOT, I repeat **DO NOT** use these numbers for your costs or markups or as an indication of what your company should be doing. I find a lot of contractors who are deceived into thinking there are some "magic" numbers floating around and when they find them they will become instantly successful on all their jobs. There are no such "magic" numbers floating around the construction industry or any other contracting industry! Each company must arrive at these figures on their own. This book should help you do that.

What this book is Not.

This book is not a startup manual for contractors. Having said that, let me emphasize the need for knowing the pitfalls of being in business. This information is as important when you are starting out as it is once you are in business - hopefully less costly! Having the knowledge of this book will help you plan your startup better and help avoid many mistakes rather than having to overcome those mistakes later.

What this book Is.

This book is divided into six sections. Four of them deal with major aspects of the construction industry and the other two are "practical experience" or "personal history" sections. Although these do not include all of the various important factors, I feel they are the most important ones needed to help succeed in the industry.

Charting the Right Course and Navigating with People.

The winds of change will never stop blowing, but with the right tack you can run as fast into the wind as with it. The need is to know where the wind is blowing from, how strong it is, and how to make the right adjustments for maximum speed under existing conditions. This section will help you and your people move together on the journey of success.

The Trouble with People and Their Performance and What to do About It

Working well with people can best be accomplished using the process known as Teamwork. Teamwork is an easy word to use, but a very hard fact to accomplish. It's like hitting a moving target. Just because an employee had the right site picture before does not mean you have the right one now. Staying on target requires constant tracking and adjustment. This section will help you track the people problems in your business and make the adjustments to stay on that target.

Taking Care of Business! Are You Running It or Is It Running You?

Are you lean and mean or fat and sassy? Running a business of any kind is not easy. No one said it would be.

Paying close attention to key principles can keep you profitable and help ensure long term success.

Growing and Staffing your Business

Does the term "Grow or Go" sound familiar? Most progressive business people have an intense desire for their business to grow and prosper. Howard Eckle has spent a lifetime helping his company do that in a major way and shares with us some well learned lessons Learning from his experience can help us avoid pitfalls in our path as we grow.

Troubleshooting the Estimating and Bidding Process

Everything, including this book, needs a logical plan to follow. Like most jobs you will do in building, you must start with an estimate. "Estimating" is one of the most frequently used words in any contracting business, but is means different things to different people. Many times, "Estimating" is confused with "Bidding," and while both words are important in building anything, they must be kept in proper perspective. You probably already know a lot about generating an estimate and bid, but this section will help you avoid some common misunderstandings.

Food for Thought

As I have moved about the country for the past several years, I have come in contact with approximately 70,000 people running or working for about 1200 companies in our industry. Many have shared ideas with me and I, in turn, share them with you. These may be quick fix ideas for problems similar to ones you have or just interesting reading. The people who shared them with me felt they were important.

True Life Stories From the Real World

While not a section by itself, these "Life Experiences" are included at the end of each section and relate to real people experiences. The stories contain subject matter that fit with concepts presented in the section. Many times, learning experiences help us know "what to do," but often, it's also telling us "what not to do." I think you'll find them interesting.

Make it Work For You!

The sections listed above are subdivided by topics. After reading this book, you may be tempted to say "Well, that's that"! However, when planning a project, I suggest you refer to the topic which seems to fit your project or problem and revisit the content. You may find it extremely helpful.

It is my sincere hope you will enjoy this book and apply some of its principles to your contracting business. You are in one of the most exciting, yet difficult, professions there is today. With careful management and continual attention to details, **you will succeed.**

Charting the Right Course and Navigating with People

Charting the
Right Course and
Navigating with People

Charting the Course

Have you ever asked yourself any of the following questions?

- Why isn't everyone in this firm on the same page?

- Why must we always appear to be "rowing upstream" against the flow of business?

- Why can't I find "good" workers who understand what needs to be done on a timely basis?

If so, then join the club! If not, then you are one of the very fortunate few who have mastered your business, or you work by yourself. These questions are very common in today's fast-paced business environment. What has gone wrong? How are you (as an entrepreneurial capitalist) supposed to deal with your customers', vendors', and employees' fast changing values and goals?

To begin with, let me say that I do not think this is a new phenomenon. Lewis & Clark, two early risk-taking explorers for commercial gain—by the way, that is the definition of an entrepreneur—must have felt the same way on their travels to the headwaters of the mightiest river in North America. They traveled with people from various backgrounds and beliefs, and encountered cultures that were vastly different from their own. They were, in fact, rowing upstream with just a compass, for there were no reliable maps for the territory they traveled. They had to chart their own course.

So, history says you are in good company. How then can we "Chart the Course" our businesses will take and ensure the same degree of success that Lewis and Clark found, while leaving a legacy as risk-taking explorers in the brave new world of the next century?

The section on page 51 gets to the heart of how to run a fast-paced business. In these first sections, we will look at how we got into this adventure and study an example of how to best handle entrepreneurial stress/tension to increase your team's effectiveness.

Let's begin by assessing what we think we know from history. Certainly, today few would argue that the only constant is change. Look around you and try to think of something that has not changed significantly in the past few years. Certain religious values and beliefs are constant, yes, but how we relate to them is always changing.

This is not to say that change is bad. Air bags are a recent change and cars of today with air bags as a standard feature are saving lives.

Change is inevitable: It is how we treat change that is important. We must learn to use change proactively and to our benefit, instead of merely reacting to how changes affect our bottom line. We, as the leaders of our firms, must learn to see change as a tool, and to use it to our firm's advantage.

Consider how businesses have handled business communication. I am certain you have a phone, but it was unique to a few businesses back in the 1920's. I would guess many of us have a Fax machine, but not many businesses did until the mid 1980's. Today, a certain number of us have cellular phones, some of which are already digital; and the pioneers among us are already "on line" with the Internet, talking with vendor and customer alike.

One word describes what is causing both the *speed* and the *magnitude* of change to accelerate so rapidly: technology! The graph on page 27 depicts the passage of time along the X axis and the body of human knowledge along the Y axis. For the first several-hundred-thousand years we were hunters and gatherers. The skills necessary to be good at it were passed on verbally from generation to generation, and an individual's source of wealth was his or her own physical strength and hunting skills. Next, as humans grew in number, we gradually became an agriculturally based society, which we maintained for several thousand years. The skills necessary to be successful were still passed on verbally from generation to generation along with a new invention—the book or written word. An individual's source of wealth now became owning land and then having farming skills. Following the Renaissance, we moved more dramatically toward becoming an industrial society, which we were for several hundred years. The skills necessary to be successful were now passed on by a shop foreman (instead of always being a family member) by both verbal instruction, reading and hands-on doing. An individual's source of wealth now became the ability to acquire capital, to buy land, buildings, and machinery to support industry.

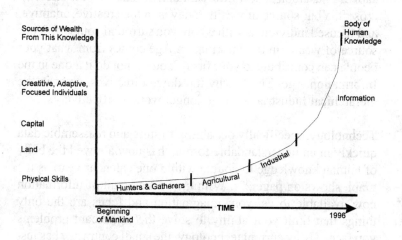

Historical graph of Human Knowledge vs. Time.

Obviously, fewer and fewer people were able to acquire this wealth. Sometime in the 1970's, we started moving into the next age of man, the Information Age. World War II had dramatically altered the balance of industrial power and, as we rebuilt most of the industrial world we set ourselves up for falling behind competitively by rebuilding those countries with the latest machinery. We have only been in the Information Age approximately two decades and we already see a shift in emphasis from the hardware which stores massive amounts of data to software and the way information is retrieved and processed. The skills necessary to be a good Information Age worker are evolving quickly and require rapid learning abilities using multiple medias. (books, videos, CD Roms or a human teacher. Imagination, creativity and vision are the are the attributes which make an effective worker in the Information Age.

Because of this, as we have seen in the last few decades, an individual's source of wealth is no longer his strength, land, or even capital—just look at how little capital Bill Gates used to start Microsoft. Your source of wealth today is in the creative, adaptive, and focused individuals with whom you surround yourself. Your source of wealth in the Information Age comes from what your people can contribute to your firm. You cannot do it alone in the Information Age. That's why top-down directives, great in the hierarchical Industrial Age, no longer work as effectively.

Technology, specifically our ability to store and reassemble data quickly in an understandable format, has now allowed the body of human knowledge to double within one calendar year, as the graph shows on page 27. With the vast amount of information now available to us, your imagination and focus are the only things that limit your ability to solve the significant problems you face. Using current technology, the small contractor has just as much processing power available to him or her as huge corporations had only ten years ago. PCs and software operating systems of today can store more data and return it to the operator in a more understandable format now than the machines could just half a decade ago. Now the balance of power shifts to the expertise of the operator!

This theory may explain why it would appear that you are always rowing up stream. But, why can't you get everyone on the same page and why don't your employees—your source of wealth—understand what needs to be done?

One part of the answer can be illustrated by a cube (see below). Which corner in the illustration—A, B, C, D, E, F, G, or H—is closest to you? Got an answer? Look again. Now ask another person, and you will get a different answer. Some people would say D, others G, still others C. It depends on what you perceive the "base" of the cube to be. If you see the "base" of the cube as being corners B, C, G, H, then D is closest to you. If you see the "base" of the cube as being corners C, D, F, H, then G is closest to you, and so on.

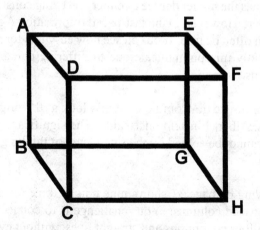

Paradigm Cube

We do not all see the same thing looking at a stationary cube. With the speed and magnitude of change happening around all of us, it is no wonder we do not all see the same thing—when we are looking at the same thing! We all have different "paradigms."

A paradigm is a map or model of the world, which we make to better understand that world. Different people map out the fast changing information around them in different ways. Our politics, cultures, and—in business—our industrial competitors all color reality. Republicans vs. Democrats, sushi as a meal vs. bait, our firm vs. their firm in industrial affiliations, are real-life examples of conflicts caused by differing points of view.

Your people think that they are making good and correct decisions in your business. Are you seeing them making inaccurate or bad decisions based on *your* perception of the same issues? Generally, experience and expertise are color perception, with whoever has the higher degree of either one being considered the more expert. However, in the fast-paced Information Age the opposite can often be true. You can see how easy it is for people to inadvertently miscommunicate due to differing paradigms and perceptions.

So, where do we go from here? Let's look at the origin of the problems. Albert Einstein once said, "The significant problems we face cannot be solved by the same level of thinking that created them."

The drawing on page 31 shows nine dots in three rows of three equally spaced columns. Your challenge is to connect all nine dots with four continuous and straight lines, without ever lifting your pen. Go ahead and give it a try. You will quickly find that you must extend your lines outside of the square in order to accomplish your task. The same is true in running your businesses today. You must extend your firm's talents and abilities beyond the "standard business square" in order to accomplish your goals. We must learn to think outside of the square, the other half of the management paradox. This requires harnessing your people's "hobby energy."

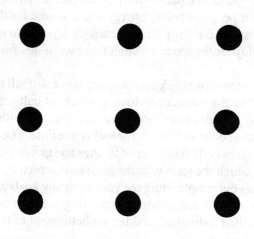

Thinking Outside of the Square

Hobby energy is the energy that anyone has to take part in their favorite hobby or activity, after they are done with workaday efforts. Have you ever noticed how excited your people get around 5:01 P.M. when they are looking forward to taking part in a softball game or bowling league, despite the fact that they worked hard all day? Suddenly, they have the energy to run, jump, and slide during a game, when just 30 minutes earlier they were putting off until tomorrow what they could have done today? That burst of energy is hobby energy, and it is generated in people who sense fulfillment in activity they enjoy.

Think of your people and their diverse talents and abilities. Now, you simply need to create the desire for them to use their hobby energy within your firm in a synergistic form to fulfill both the company's desires and the employees' desires. If done properly, everyone benefits from the combined talents of the firm—all focused to meet your customers' needs and expectations. Today we must expand our paradigm of management and attempt to find out our people's hobby energy and a source for the fulfillment of it within our firm. It is not enough anymore to merely reward employees for accomplishing tasks we assign them.

Does this mean changing your company into a softball team? No way! But take a moment to look at what softball really is—a team sport, played with an individual focus; that is, with only one batter at bat or one fielder with the ball at one time. The game has defined rules; a certain number of innings and number of outs per inning, in which the team with the most runs scored by individuals wins. If your people could see your company goals as they see batting for a home run to win a game, they would want to each contribute their individual best for the betterment of the team.

How would they do this? Staying within the rules, they would perform their tasks within the days allocated for them with all their competitive spirit so that the company has a winning season, and they would share in the pride of that accomplishment. Getting your people to work in this way is not an easy task, but one with an enormous payoff in an age when your people are your source of wealth.

Change will not occur overnight, and a new understanding of managing human behavior will not be accomplished by reading one book! Because we are the first generation of managers to grow up in the Information Age we do not have the luxury of having managerial insight passed on to us by previous generations. As a matter of fact, just the opposite is true. We are the first generation who will probably evolve into a second age of man in fewer years than the average business career lasts. Because of this we tend to manage from our historical background and education instead of managing from a proactive future orientation.

In order to be more effective, we must learn to balance the following:

- *Change* with *history*
- *The personality types within the team*
- *Effects of personality types* with *a persons job type*
- *Company goals* and *expectations with individual's goals and expectations*
- *Flexibility in our job descriptions* with *stability in our business*

In an age when the only constant is change, we must be acutely aware of each of our people's concern over the change they see in the business versus the history of a firm. Older employees, meaning those who have been with your firm the longest, tend to be more concerned with the effects of change. Younger employees, on the other hand, are more concerned with who has seniority in front of them and what is their potential of becoming a long-term employee with the firm.

The balance between flexibility and stability is another issue you must learn to monitor within your people. Some people are drawn to hierarchy and do not want to get involved in decision making. Rather, they prefer to take commands and instructions, then carry them out. Other employees need flexibility; they need that decision-making freedom to thrive. However, you must understand that with flexibility comes ambiguity in how a specific task will be accomplished. No two people accomplish a task in the same way because of their individual paradigms, perceptions, experience, and expertise.

Many books and theories have been written about understanding different personality types and behavioral styles; Myers-Briggs, Carlson, and Erikson, to name just a few. It is not within the scope of this section to be able to discuss all of these personality types and behavioral styles in detail. As humans we are a complex mixture of all these elements. It is our ability to know and use this knowledge in staffing our firms that will help us survive into the next century.

Generally, personality types are based on four major criteria: people interaction skills; learning styles; emotional formats; and, life ordering preferences.

Some people gain energy by being around other people; while other people are recharged by being alone. You can keep this in mind as you place extroverts and introverts within your firm.

Learning styles differ according to whether a person learns from facts, empirical experimentation, or theory as opposed to those who learn intuitively or through hands-on or kinesthetic learning. You can know that bookkeepers, accountants and estimators tend to lean toward facts, theory, formulas while job foremen and field workers tend to be more intuitive in their learning style.

Emotional formats tend to be the easiest traits to discern in that people are either thinkers like Star Trek's Mr. Spock, or feelers who decide with their hearts. A good balance between the two is ideal within a firm.

Finally, a person orders his or her life in different ways. The terms "player" and "judge" say it all. "Judges" prefer a very orderly and regimented life, while "players" may order their life somewhat, but will drop the plan in a flash if the opportunity for fun should suddenly pop up. These people tend to be very spontaneous, and may appear to be happy-go-lucky in their approach to life. Keep in mind that too many employees with traits that are too strong in either direction will make for a miserable firm. A job foreman with high "judging" traits will proceed dead ahead, just to stay on schedule, while a project manager with high "player" traits will start any project that appears to be fun, but may not be a strong finisher. You have to find a balance in your company.

Behavioral styles are personality traits brought to life in the working world. The four major styles are:

D·· Drivers - High ego strength, loves change, bottom-line oriented, strong competitors

I ··· Influencers - Strong ego, use change, get things done by influencing people, strong need to be liked

S··· Steadies - Low ego strength, hate change, are the people who get things done, strong desire to please

C·· Calculators - Firm ego strength, measure change, count and calculate everything, strong desire to look at all the possibilities

I am sure you can see some part of yourself and all of your people in all of these personality types and behavioral styles. Note that we tend to adopt different styles somewhat at different times and places of our lives.

One word of caution: As the CEOs of our firms, we tend to do two things which can be counterproductive: We tend to hire "ourselves"—people with similar styles and traits, because we can just relate to them. Be careful, a firm with too many people of one style can really narrow your effectiveness. Also, people with opposite styles and traits can be intriguing to you in that somehow you know they have something you are missing. The difficulty comes in not understanding why people behave the way they do under the pressures of business. Opposites attract, but if you do not learn to appreciate the complimentary nature of your styles and traits, you are in for a long battle.

Is it any wonder then, that with the business world changing so fast and with all these people in your firm having such differing paradigms, perceptions and personality types and behavioral styles, that we feel like our people are not on the same page with us or understand what needs to be done? What are we to do? Is there an answer?

One of the answers lies in the "Fulcrum Point" shown in the following drawing:

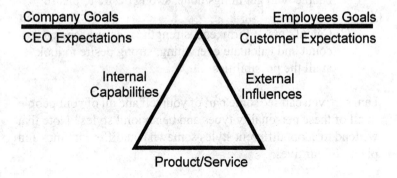

The Fulcrum Point

What you have been feeling is a struggle to balance your expectations for the firm and the goals that are beneficial for it with your people's individual goals—often not clearly known or expressed—and your customers' changing expectations of you. You must balance these differing paradigms, functioning literally on a teeter-totter. Your base unit is the product or service you provide, along with your assets, people, systems, culture, and constantly changing external influences: social, economic, political, technological.

The "Fulcrum Point" is where all of this energy comes together! It is the point with the greatest amount of stress and friction, and the point of the greatest amount of freedom and flexibility. Welcome to the definition of entrepreneurial ownership. It is the point of paradox!

Paradox is often seen as bad in today's business world. There is this tension or stress between growth and stability, change versus tradition. However, what we must learn to do is to accept some degree of tension as being natural and actually beneficial for our firms. It is what stretches us and makes us grow! It is finding the right degree or "Fulcrum Point" of tension that is critical to success in today's fast-paced world. You must learn your tolerance and your people's tolerance levels for this stress and then work to keep your firm at that balancing point!

The other answer comes only after you have decided to try to live with a reduced amount of tension in your managerial life. For this, we must look at another occupation that deals with a constantly changing environment, has high degrees of risk, and ultimately provides the type of excitement that draws people to take up the challenge. Let's go whitewater rafting!

Go With the Flow

Whitewater rafting is a fast growing outdoor activity with thousands of enthusiasts across the country. It is a combination of individual effort, guided instruction, and team spirit. There is nothing quite like the satisfaction one gets after successfully running a daunting set of rapids with your raftmates. The price of failure is high; the thrill of victory is exhilarating—just like life in the entrepreneurial business world!

In order to enjoy whitewater rafting you need three elements: a raft, an elevation drop, and a river. The raft, as we will shortly see, can easily be related to the work team we find in our businesses. The elevation drop can be compared to the environmental changes that impact our businesses—that impact all businesses, to varying degrees. The river can be compared to your business in that rivers and businesses are created in the same way, grow in similar fashion, create unique "ecosystems" around themselves, and certainly provide the ride of your life if you are brave enough to ride them.

A river is created in its headwaters by runoff from melting snow. The runoff follows the path of least resistance into valleys to create brooks; several brooks merge to become creeks; creeks join together, with further elevation loss, to become streams. Streams run from mountains to become small rivers in valleys first; then become progressively larger rivers as they flow to the ocean.

In the same way, businesses start with the creation of an idea or the thought of doing something a better way. Soon, the idea progresses beyond the thinker and attracts other people. These people will have a deep sense of loyalty to the company as long as they remain employees. The idea grows and starts to produce a profit, which brings more employees and "units of activity" such as vehicles, assets, etc.

Just as a river is made up of many different flows, a company starts to develop many different goals, as the number of different people needed to facilitate its growth expands. In the worst case scenario, if a business grows too fast it is like a river overrunning its banks, and its growth can become destructive to both the company and the surrounding environment.

Brooks, creeks, streams, and rivers each have differing ecosystems around them, based on the age of the plants and variety of animals and fish which inhabit them. As companies grow in size and different phases, they tend to have differing strengths and challenges, based on employee and customer expectations. A small startup firm with only a few employees and assets cannot tolerate any "evaporation" of assets, and the major concern of its managers is survival. Some companies choose to remain this size for the life of the company. Others—some by conscious choice—continue to grow and start to accumulate more employees and assets. These firms are busy with the proper acquisition of assets. Their major concern is fueling their continued rapid growth. Many firms enter this stage, but few mature to move on to the next stage. Here's why: Something happens in the environment—a shift in consumer demand, a split up of personnel, or a downturn in the economy—and, suddenly, the rate of rapid revenue growth slows.

Unfortunately, during this second stage, the rate of growth of expenditures, such as payroll, overhead, expenses, etc., tend to exceed what "standard" business expenditures would be. If anything slows the superior rate of revenue growth and the owner cannot or will not throttle back on the expense lever fast enough, the firm is forced to eat up its skimpy asset base and take on debt—or worse!

If a firm is lucky enough, or well managed enough, it will progress on to the next stage of growth. This is where the rate of growth slows and most firms begin to departmentalize or create new flows of revenue growth within themselves. These firms start to focus on the prioritization or allocation of employees and assets to maximize the firm's return on those assets, instead of just adding more employees' assets to the expense base. They now start to focus on the depreciation and upgrading of assets. Their major concern is now maximizing return on assets used. It is possible to handle all these challenges in one business career, however most people are better suited for life in one particular phase of growth or type of company. It is a rare individual who has the capacity to instinctively handle all the challenges faced in each of these different phases of growth.

Just as a river is made up of hundreds of different flows, your business is not just one constant flow of commerce from the original idea. Your business has many different flows of commerce; some good, profitable jobs, like the main channel in a river, and others like a pool, break-even jobs that give you a place to rest but not necessarily contributing to the company's goals. Reverse flows, unprofitable jobs, actually flow against the main channel and create turbulence within the river, which causes erosion of the banks, in more ways than one!

One of the keys to running a successful business is to be able to identify which phase each of the different flows of commerce within your business you are in. If you can maximize your good flows—the main channels—while minimizing the turbulence—the unprofitable jobs—within your business, your business will flow more smoothly.

All rivers have obstacles in them that divert and alter the flow of water around them. In your business, these rocks tend to divert energy and resources away from the pursuit of company goals. Rocks can be large or small, and tend to be elements like weather, economic factors, personnel issues, technology advancements, and/or competitive firms. Most business owners just hope that the rocks will dissolve or that enough revenue will flow through the river to cover the rocks.

Those of you who have been whitewater rafting know how foolish it is to think that a river will be easier to run in "high water." More water brings on more thrills and far more spectacular spills. It is much more effective and practical to learn to guide your business through these obstacles than to just hope the revenue will rise over them. Once you have the confidence that your raft can handle any level of problem you face in this stretch of the river, you are in much better position to handle the problems which loom down stream!

If business is like a river with flows, currents, and rocks, then how should we attempt to maximize our "return on energy"—company profits, heightened reputation, and personal satisfaction—that we put into it, let alone enjoy personal fulfillment or employee enrichment?

I live in Colorado where we have world-class whitewater river rafting, which is enjoyed by a growing number of enthusiasts each year. What is it that attracts these people to this type of adventure vacation? Aren't vacations for lounging by the river, not running directly through it? Certainly people will always seek to find new ways to fulfill their ever-changing cup of desire. Whitewater rafting provides a sense of accomplishment, thrill, and hard individual effort that leads to team success. After rafting a river, you will never look at a stretch of river the same way. You also develop a greater understanding and appreciation for the skills required to successfully run ever-changing river conditions. Aren't these all attitudes we would want our employees to feel working within our firm? And, is it any wonder that when our key employees are not able to gain this sense of fulfillment from working for us that we lose them or worse yet, they become our new competitor?

If you decide you want to ride the river, you may as well apply the "DIRTFT" (Do It Right The First Time) principle to it, as taught by Bruce Merrifield. Life is too short to keep redoing "WIPEOUTS" (Well-Intentioned, Poorly Executed Opportunities). Besides, once you "WIPEOUT," or your or your people's expectations are not met, you will simply think "I told you so," and go right back to your old time-tested habits.

Wipeout has been going on since the days of the hunter/gatherers. If you cannot properly farm, just kill the animals that are eating your crops. If you cannot properly run a machine in the big city, just move back to the farm. If you cannot properly learn to utilize a computer or any form of technology, just look for a high paying assembly line job. History shows us that despite the wipeout effect on any single individual, we as a culture will, in fact, move on to the next stage of mankind.

So how do we go "DIRTFT principle" rafting? First of all, never try to whitewater raft on your own! Every year many people die in the river because they think, "How hard can this be?" If it is your first time trying any "wipe out" procedure, always seek professional help, on the river and in your business! After locating a reputable guide service, the guides will assign you to a raft of up to eight people. Normally these people have various abilities and differing emotions about taking part in rafting. (Sound like your business yet?) It is vitally important to get properly suited up for your trip with a wet suit and life vest, in order to maximize your safety and comfort. Then it is time to meet with your guide to go over safety precautions, rowing instructions and what to do when everything goes wrong and you have to get a "swimmer" back in the raft.

Knowing how to get a swimmer back in the raft is vitally important, because there should be no confusion at all on how to "re-staff" the boat during an emergency. Think how often we cover "swimmer rescue" with our work team. Don't we just have a warranty crew to fix all those mistakes?

Next you move your raft down to the edge of the river and go over a few last-minute emergency safety signals and real-life rowing commands. Then you "put in" and feel that first rush of adrenaline when your skin comes in contact with the ice cold water. Remember that feeling when you first started your business?

During the easy stretch of the river, your guide will start you with a few basic maneuvers: right turn, left turn, high side. Never put a new crew into a difficult stretch of the river first! The guide will usually make a few positional changes due to various individual's "stroke power"—their ability to listen, willingness to work hard, and their experience and drive—so that one side of the raft is not exceedingly stronger than the other. Imbalance of power within the crew would cause lots of trouble in a rapid.

Once the guide feels comfortable with his crew's positioning, you will move quickly to your first set of rapids. You always *hear* the rapids first; often you cannot see them because the river is falling away from you. The guide will quickly scan the crew for anyone who is going to "freeze up" and will get them into the middle of the raft. If they are not going to contribute, he'll at least get them out of the way of those who are.

After going through the first set of rapids, it is important for the team to celebrate success, then bail out any excess water taken on so that the raft is more maneuverable for the next set of rapids. Remember to celebrate success with your crew in your business, and then go over efficiency improvements for the next job.

Often the guide will make one more set of positional changes so that the balance of the day is both safe and enjoyable for everyone. There can be some grousing over positions, but usually people find they end up enjoying the position the guide put them in, even though they felt they would like something else prior to running the river. If the rafters continue to listen to their guide's instructions and work hard in unison with their raftmates, they will very quickly gain a team spirit as the team tackles its first few sets of rapids. The team will begin to anticipate the guide's instructions and start to develop "synergistic reactivity," where each rafter starts to sense when they will receive an instruction and what that instruction will be, and react in a unified fashion to the command.

I have been on rafts with people that know *what* to do and *want* to do it, but get in each other's way doing it, and cannot seem to get the hang of knowing *how* to perform the command. To put it mildly, there is a lot of bailing of water on those trips. However, the feeling of great satisfaction and power that you get when you are on a raft where your teammates know what to *do*, they want to *do* it, and they synchronize their efforts so that the how to *do* is in harmony is terrific. The guide will notice this and start to challenge the raft by taking "unique" lines through the rapids or even taking them backwards, which is truly a thrill for all the rafters.

It seems that no raft trip is complete without the raft experiencing "swimmers," people who are knocked out of the raft and into the rapids. Rocks are what create the channels of water in a set of rapids. The raft is constantly bumping into them, even with the best set of teammates. Some guides even use the technique of hitting strategic rocks with raft, halfway through a set of rapids, to set up the raft for its next line through the remaining part of the rapids. Upon hitting a rock particularly hard, you or a teammate find yourself flying through the air and out of the raft. This is when preparation and "synergistic reactivity" really kick in. The guide does not have the time to tell you what to do in this potentially life-threatening situation nor would you be able to hear him if you were under water.

The preparation comes in the form of remembering what to do from the safety talks you had prior to rafting. Your first instinct as a "swimmer" is to want to get back in the raft. However, in a set of rapids, if you cannot get into the boat cleanly, you can very easily be pinned between a rock and the raft which *will* cause injury. The guide will decide if you can get into the raft in time or if you will do the balance of the rapids as a floating object! This is where "synergistic reactivity" comes into play. If the guide knows that each member of the raft can perform not only *their* duties but also the *new* duties now required of them by this emergency situation, it will greatly lengthen the window of time for recovering swimmers in a set of rapids. Someone will have to lift the swimmer into the raft, while others continue to stroke per the guide's instructions, for there will be still more rocks to negotiate in the rapids.

If by chance, you as a swimmer are forced to run the rapids without the aid of a raft, you should position yourself in the whitewater swimming position, on your back, arms out, feet downstream, and breathe when you can. You need to get to shore quickly, but safely! Oftentimes under rapids are log jams of trees, branches, etc., which are extremely dangerous. They are called "strainers." Large trees have become lodged between larger rocks and are straining smaller debris out of the river—in this case, you. You will be sucked under the "friendly" looking log. This is not good! Remembering your safety training preparation, you need to steer clear of the log, which appears to be your quickest way out of a bad situation. Swim to the rocky shore and your raft will soon come to pick you up. All you have then is the story of a lifetime from your rafting experience.

The other event that seems to occur with a fair amount of frequency while rafting is the "high side" situation. You are running a rapid, you hit a rock and it turns you side ways! Rafts are not designed to go down rivers sideways; they are impossible to steer, and actually not very stable in that position. If your raft hits a rock while traveling side ways, it tends to ride up on that rock, which lowers the upstream pontoon, which fills the raft instantly with river water, which flings all of the rafters out of the raft. This all happens in a matter of seconds. Now you can forget "synergistic reactivity" because what you have is eight "swimmers" and a raft stuck on a rock. Everyone needs to get to shore as quickly and safely as possible and you will need another raft within your party to bring your raft to shore.

One way to avoid this mess is for the guide to instantly call for a "high side" when the sideways raft initially hits the rock. "High side" is when everyone in the raft instantly jumps to the "high side" of the raft, thus taking all the weight off of the lower pontoon and hopefully raising it above the onrushing river water. Eventually the river will sweep the raft around the rock and everyone needs to jump back into their position to continue through the rapids. Only diligent preparation and quick reaction time can ensure your team will live to tell their high-side story—dry.

As you can see, there is a lot more to whitewater rafting than meets the eye, just as there is more to your business. But if you prepare ahead of time, have good equipment, position people effectively, communicate properly, and start to develop synergistic activity, you can enjoy a whitewater or business experience!

Take a moment to review the comparisons between river rafting and running your business. In the space between each item, jot down a few action items for you to work on with your firm to make running your business more enjoyable and productive:

River Has:	Business Has:
Action item:	
DIRTFT · · · · · · · · · · · · · · · · ·	Do It Right The First Time
Action item:	
Flow · · · · · · · · · · · · · · · · · ·	Revenue Streams
Action item:	
Main Channels · · · · · · · · · · · · · ·	Core business practices, tend to be most profitable
Action item:	
Side Streams · · · · · · · · · · · · · ·	New opportunities and/or threats
Action item:	
Pools · · · · · · · · · · · · · · · · · ·	Seasonal, "break-even" jobs
Action item:	
Reverse flows, turbulence · · · · · ·	Misdirected efforts, "jobs from hell"
Action item:	
Evaporation · · · · · · · · · · · · · · ·	Mishandled assets; profit "leaks"
Action item:	
Rocks · · · · · · · · · · · · · · · · · ·	Obstacles to business; competitors, poor economy, regulations, personnel issues
Action item:	
Flooding · · · · · · · · · · · · · · · · ·	Destructive growth; growth with out control
Action item:	
Bank Erosion · · · · · · · · · · · · · ·	Pocketbook erosion
Action item:	
Surrounding Ecosystems · · · · · · ·	Phases of Business Life Cycle
Action item:	
Guided instruction of individual effort, which leads team tosuccess beyond their normal sense of accomplishment	Guided instruction of individual effort which leads team to success/adventure beyond normal sense of accomplishment
Action item:	
Main desire is to maximize fun while being safe	Main desire is to maximize return while growing sense of fulfillment for all

Rafting Has	Business Has:

Action item:
W.I.P.E.O.U.T. · · · · · · · · · · · · Well-Intentioned Poorly
Executed
Opportunities Utilizing Talent

Action item:
Raft of six to eight people · · · · · · Project manager, foreman, work crew

Action Item:
Elevation Drop, which · · · · · · · · Environmental changes that
causes rapids impact all firms to some degree

Action item:
Rafters with differing · · · · · · · · · Employees with different goals
abilities and desires and objectives than CEO

Action item:
Rafters with differing · · · · · · · · · Employees with differing
"Stroke Power" abilities, expertise, and desire to
produce work

Action item:
Guide's ability to develop · · · · · · CEO's ability to develop
raft "synergistic reactivity" employee "synergistic reactivity"
with work teams

Action item:
Wet suits & life vest to · · · · · · · · Company mission statement and
maximize enjoyment job description to maximize
and safety effectiveness and fulfillment

Action item:
Safety hand signals · · · · · · · · · · Previously agreed upon critical
(because you cannot hear recovery plan
the guide in the rapids)

Action item:
"Putting in" · · · · · · · · · · · · · · · · Go for it

Action item:
Practice Basic Maneuvers · · · · · · Get people to see how job
after "putting in" description "works" in real life.

Action item:
"lines" through a rapid · · · · · · · · Putting tactical action plans in
guide'sstrategy for running place to hit strategic goals
a rapid

Rafting Has	Business Has:
Action item:	
Guide makes positional changes to maximize "stroke power"	Owner must balance work team for efficiency & effectiveness
Action item:	
Run your first set of rapids	Work on first project - encounter "rocks" and obstacles
Action item:	
Guide makes further adjustments based on who handles pressure and who "freezes up"	CEO must balance work teams to maximize production and follow-up
Action item:	
"Swimmers" people who are thrown from the raft by rock or raftmates	Personnel issues which slow the production of the team. Culprit is often miscommunication.
Action item:	
"Strainers" - Log jams in the river which look inviting but are dangerous	The easy way out of a difficult situation instead of solving the problem.
Action item:	
"High side" command given in dire emergency	"Damage Control" - review emergency procedures before they happen so employees with cool heads can handle situation.

Synergistic Goal Orienteering

Here is a process of developing a step-by-step action plan that will focus your key people's "hobby energy" on mutually agreed upon, prioritized goals. It is the art of making dreams come true, through structure and collaboration: synergistic goal orienteering. Before discussing the step-by-step process, let us first take a look at the big picture of how this orienteering works.

Synergy is defined as a process where the whole is greater than the sum of the parts. To follow our river analogy, the power of a mighty river far exceeds that of the sum of hundreds of streams. And the force created by your people's hobby energy, a powerful force in itself, focused on common goals, far exceeds the force of the sum total of their individual talents. This is also exemplified by the image of a laser as seen in the following drawing:

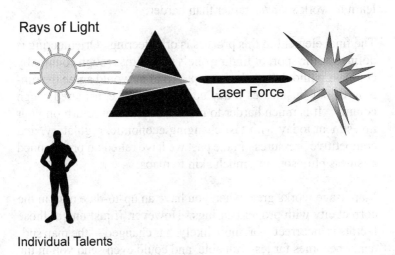

Focus your people's talents on common goals.

Your people's talents would be similar to the individual beams of light focused by a lens. The lens is your firm's action plan, which has been mutually agreed upon and prioritized into strategic importance. With that type of focus, your people's energy could not help but strengthen synergistically into a force well beyond the combination of individual talents. That is the good news. The bad news is that in today's business world you need this synergy just to stay even with the effects of rapid change!

The next element is a goal. A goal is defined as an objective with a purpose. Through the synergistic goal orienteering process, you will not change your people's goals, you will only learn to focus them in a more concerted effort. Lasers do not change the light waves into another form of energy, rather they simply focus a wide spectrum of light waves into a narrow, powerful beam. This is far more powerful than any one individual light wave. This is a hard paradigm to learn for the average CEO. What has worked for you in the past is to work *harder*. Now you must learn to work *smarter* rather than harder.

The final element of this process is orienteering. Orienteering is defined as the sport of finding one's way across rough country by compass without the aid of a map. All of us would agree the running of our businesses today could easily be considered rough country. It is much harder to make a substantial return on your investment today with fast changing economic, regulatory, and competitive pressures. In the past we have relied on pre-planned business philosophies, much akin to maps.

Map usage works great when you have an up-to-date map in the correct city with proper bearings. However, if just one of those factors is incorrect—or, more likely, has changed— the map suddenly becomes far less valuable, and could even send you in the wrong direction. No one is in more trouble than someone who thinks they are moving in the correct direction, for the right reason, when neither is the case. This is the same reason that lost people walk in large circles; they have a strong or dominant leg, which takes a slightly longer stride with each series of steps. Thus, the lost person with good intentions, gets increasingly lost with each new effort put forth.

If maps are becoming increasingly suspicious, CEOs must learn to use a compass to help guide them to their destination. A compass is a device with an electromagnetic needle that points to North, thus allowing us to get our bearings no matter where we find ourselves. North is a constant, it is one of a few things in nature that does not change! We can learn to use its power to help us get our bearings in any business situation. If we learn to use our people's focused energies to achieve our strategic goals, we will hit those goals more often than not.

The promised land, when we purposely achieve our desired objectives, is our ultimate goal. What we hope to learn is how to use the power of human compasses to help us get our bearings straight in an ever-changing environment. We should prepare for the worst, expect to hit a rock or two, and celebrate our every success. Then we can chart the course our firms will take through the rough country of our industry, toward the goal of synergistic need fulfillment: satisfaction for our customers, profit for the company, and fulfillment for our people's life goals.

Success is a Journey, Not a Destination

We have come far in a short time. We now have an understanding of why our people, who are so important to us, don't always seem to have the same goals as our company. We have come to an understanding of the significant role that change will play in our lives and how we can use that change to our advantage. We have discussed synergistic reactivity and how it helps our firms cope with the rocks in the river of commerce. And, we have looked briefly at goal orienteering, the process of focusing your key people's hobby energy onto mutually agreed upon, prioritized, strategic goals.

If we are to be able to generate the power of a laser from our people's light waves, then we must use the proper lens. In this section we will discuss the real-life process of developing that lens in our firm, so that no matter who works in our firm, we can always operate at peak effectiveness. It is one thing to decide to chart the course our futures will take, and quite another to find a better way towards a greater result!

The first thing you need to do is plan and execute a company retreat with your key people. You know who they are—the 20 percent of the people who get you 80 percent of the results. Why a retreat? Well, just as your professional river rafting guide took plenty of time to go over safety precautions, rowing instructions, and individual positioning prior to getting in the river, we must do the same thing. I know far too many firms that have a three- or four-hour meeting on a slow afternoon and call that strategic planning. You need to get your key people away from ringing phones, "forest fires," and other urgent business activities. (Notice I did not say *important* but *urgent*. How many times have you spent hours in a day doing urgent things, and then at night recall all the important things you now need to get done tomorrow? We have all been there, done that!)

Think of the message you send to your key people by involving them in a planning retreat. "I value your opinion and I need your help to accomplish our goals." This message will draw hobby energy from virtually every employee. Timing is critical; you need to plan *prior* to your busy season, yet during a time when scares from the last poorly executed "high side" flipped your business raft. A Friday-to-Saturday retreat seems to work well: one day of company time, one day of personal time. You will have two days for critical open communication and strategic goal orienteering. Be sure to give your people plenty of advanced notice so they can take care of family issues. You want every one there!

Location is also important in that you want to establish an enlightened atmosphere and "get away from the office" feel. Virtually every city in North America has a nice destination facility near it.

You may also want to invite the spouses of your key people along, to really make a statement to them about how much you care. The spouses already have a definitive opinion about you and your firm, and this is a great way to say thank you to them in an unexpected way.

Many of you have already kissed this idea off because you are trying to figure out what this is going to cost the company. Well, how about talking with your two or three top suppliers and getting them to help co-op your strategic planning meeting? Say you are in the landscaping industry, you could contact your nursery, sod farm, and irrigation supply house about helping to defray some or all of the cost. This works especially well when none of them are competitors and the suppliers feel like they can each take full credit for their assistance. After all, why wouldn't a distributor want to help you with your strategic planning costs? If you become a stronger customer in the future and show them you are following a professional model for the meeting, isn't that a win-win scenario for everyone involved? Do you have any idea how much more it costs to find a new customer instead of just growing an existing good account? Of course you do, you are in business too!

Deciding who to invite within your firm as key employees is the third critical thing to think about, and perhaps the most important component of a successful strategic planning retreat. You can always go with the seniority model and just take the people who have been with you the longest, or you can go with the payroll model and just take the top 20 percent of your payroll. However, I suggest you take the time to sit down and think about irreplaceability. No *one* person is irreplaceable in any firm, even you the CEO can be replaced by another CEO and different capital. But each person has a "cost to the firm" associated with replacing them. Think what it would cost to replace a key employee who would take with them their customer contacts, industry knowledge, and institutional experience. Each employee has some cost. Take the time to rank them honestly by their replacement cost. Then, invite the people from the top of your list to the planning retreat. You may find that the irreplaceability list is very different from the seniority or payroll list. If you have a properly executed planning retreat you will definitely hear this!

I would also recommend you invite two additional people to play two very important roles in your meeting to maximize its success. These two people should be company outsiders, so they will be impartial to the discussions that will go on. The first role is that of facilitator, a person who will guide the group through its discussion and help keep it on track. Someone with strong listening and interpersonal skills is highly recommended, and as with rafting, if you hire a professional you greatly increase the odds you will have a safer and more enjoyable planning retreat.

The second critical role is that of chairman of the board. You need to have someone who will play the role of board of directors and constantly keep asking the group "How will what you are discussing bring a better bottom line to our firm?" You will be discussing many issues that do not have an immediate product or service implication to your firm and it is this person's responsibility to help the participants keep realistic company profits in mind during the discussion. Professional consultants, industry associates, and vendors are just three places to look to find these key people.

Once you have decided when, where, and who to invite, you need to put all the details of the retreat together. A good meeting planner can do this for you easily.

The last thing you need to do prior to the planning retreat is send out to your key people a pre-retreat survey. The survey will help your people start to focus their thoughts and opinions, especially if you have never done this before. The survey should include:

1. What are your expectations for the retreat?

2. What markets should we pursue in the next three years?

3. What products/services should our firm offer in the next three years?

4. What one thing should we be doing by the year 2000?

In addition, you should also ask your people to write down what they think your firm's current strengths, weaknesses, opportunities and threats are today. Remember, there are no right or wrong answers. You need to see how everyone views your firm now before you start to plan for the future. The results from the pre-retreat survey will give you a strong foundation to build from for the balance of the retreat.

Once the retreat starts, in order to maximize the amount of honest and open communication, your facilitator needs to discuss some retreat guidelines. This "rowing instructions and safety talk" session will improve everyone's ability to feel like they contributed to the ride. In the rowing instructions the facilitator needs to discuss each person's role during the planning retreat.

The CEO will play the role of CEO during all discussions, but he or she must be much more willing to discuss their rationale in reaching a decision with the group than they would be in the office. How often have you as the owner said that you wished your people would *think* more like an owner, in conducting company business? How are your people supposed to learn what your driving motivations and defining rationale are in running your business, without you sharing it with them?

Your key people's role will be that of stakeholders in your firm. All firms have four different stakeholders within them. Each of these different types of stakeholders have a varying stake or sense of caring about your firm. The first stakeholder is the CEO/stockholder of the firm. The business is his or her baby, it was his or her idea, and the CEO has the most to lose; both financially and psychologically. The second set of stakeholders within a firm are the employees. Some of them have a huge stake or sense of caring for the firm, and others could not care less. The more you can create a sense of them giving the firm direction and energy within your people, the more you will see this set of stakeholders care about the firm's growth and profitability.

The third set of stakeholders within a firm are the suppliers/vendors. Traditional hierarchical business practice has looked at your customers almost as obstacles to do business around. A sort of competition exists, in which you try to do as much business with them without having to concede major gross margin concessions or pick up extra service costs in doing business with them. Normally, your biggest customers in volume are not your most profitable customers to do business with, because of this competition.

The competitive change of the 90's has brought on a new need for mutual business partners to collaborate to insure your mutual profitability. You need to work with your suppliers to help drive down costs to both of you as you do business with each other, so that you can offer the best overall value to your end customer. The growth of big do-it-yourself retailers selling directly to the end customer at the contractor's expense is a classic example of this phenomenon in the 90's!

The final set of stakeholders within a firm are your customers and the community at large. People need your business, as everyone can attest who has ever tried to go back to a defunked business to get a problem fixed or has lived through a severe economic downturn, and can no longer find a local source for what they need. Long term, they are much better off and will get their needs filled more quickly by a satisfied supplier. Your people need to guard the company's "stake" as you talk about future products or markets to pursue, which will help you all achieve the firm's goals for growth.

You will need to assign one person the role of "scribe." This is someone who can accurately write down all the major points of discussion and pencil in "authorship" so that later, after the retreat, when you are starting to implement your plan, the group can remember who said what and why. The identification of "champions of an idea" is critical to drawing out your key people's hobby energy towards accomplishing new company goals. Certainly, you can see the importance of having an accurate and detail oriented "scribe" for the follow-up success of the planning retreat.

The last two roles of facilitator and Board of Directors have already been discussed. Hopefully, now you can see more clearly how these two important roles will fit into achieving the maximum amount of impact on your firm from your planning retreat. The benefit of getting two "outside influences" to help your firm see some of its damaging paradigms far outweighs any additional cost. Remember, you can approach one or more of your key "collaborative vendors" for both or either of these roles in addition to helping offset the actual retreat costs. If they are truly a valued stakeholder in your firm, they would be more than happy to assist you in any way. If all of these people play their roles to the limit of their abilities, you will maximize your firm's "stroke power" both during and immediately following your strategic planning retreat.

Your firm is not quite ready to "put in" to a successful planning retreat. The facilitator still needs to go over the safety talk. You know the "what to do if things get out of hand" chat. If we cover these important facts now, we increase the odds that we will not create a "swimmer" from our planning retreat. There are five major safety rules to a successful retreat:

1. Understanding cubes

2. No monkeys

3. Drawing a flag

4. No cows

5. Ears vs. tongue

Let's look at each one in turn.

1. *Understanding cubes* is all about getting your key people to understand the importance of perception in

your firm. Remember the Paradigm Cube on page 29? The corner that is closest to you depends on what you think the base is. Your people will see differing strengths, weaknesses, opportunities and threats (SWOTs). Some of your people's strengths will be seen as a weakness by others in the firm. During the SWOT analysis we are not trying to *fix blame*, only move toward *fixing the problems*.

2. *No monkeys*. Problems are opportunities in disguise. As your group gets into various discussions on SWOT's or new markets to pursue, one person in your firm cannot continue to say that they cannot accomplish a certain goal because someone else in the firm will not support it. That is putting a monkey on someone else's back and will do nothing toward helping the firm achieve its goals. I use a toy monkey for the group to throw around when they hear someone putting a monkey on another's back. It keeps it light hearted, but still gets the point across.

3. *Drawing a flag*. Just as certain infractions of the rules in football draw a flag from one of the officials; so too must your group and the facilitator throw false flags at you or your key people who are not playing by the rules. You need honest and open communication during your planning retreat, not hierarchy or tenure! If one of your people will open up and tell you what they really think, you need to respect that instead of drawing an "unsportsman-like-conduct" penalty.

4. *No cows*. Your group must agree to discuss openly all sacred cows and false idols that your firm has. Every company has them. Your newer employees do not understand them and the older employees will not talk about them for fear of retribution. You must be willing

to discuss these sacred cows within your firm if you are to move onto the next phase of growth your firm is facing.

5. *Ears vs. tongue.* Everyone in your planning retreat must practice active listening skills so that they can truly hear what others are saying. Active listening is when you want to hear what others say; like when you get to discuss your hobby with any one. As humans, we can think five times faster than anyone can speak, but instead of preparing our next rebuttal we need to focus on truly hearing where the speaker is coming from. A good facilitator will have some active listening skills activities to share with the group so that everyone understands the ultimate purpose of active listening!

All five safety rules, if properly implemented, will help key people from getting their feelings hurt early in the process and checking out from the group, thus diminishing the overall effectiveness of the planning retreat. If you cover the rules up front and take a lighthearted approach during the meeting, you will find you have a lot more time for constructive conversation than you do at the office!

The last thing you need to do with your key people right before you "put in" is to review their retreat expectations with them. The facilitator—not the CEO—should draw this information out of each participant. The scribe needs to record it, so that at the end of the planning retreat you can revisit the expectations to see how you have done in meeting them. It will dramatically help the group to hear from each person what their retreat expectations were. Intercompany paradigms and individual perceptions will become very evident through these expectations. We must know as a group who is looking at the firm through what shade of glasses.

Now, you are ready to "put in" to your retreat. Good luck and may your retreat—and your adventure in business—be an exciting and profitable ride!

Now that we have discussed the roles that each individual will play, agreed on the rules of safe conduct, and given each key employee an opportunity to discuss their planning retreat expectations. It's time to put our raft into the river of strategic planning. We must initially run through a few quick commands to determine each participant's stroke power and position people accordingly in the raft.

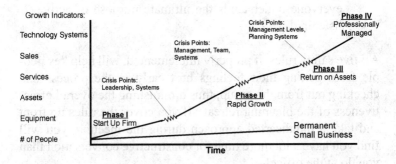

Phases of Growth

(author unknown)

You can accomplish this in the planning retreat by reviewing the *Phases of Growth Model* (above) and discussing each of your people's SWOT analyses. In the section on *Go With the Flow* (page 39) we discussed briefly the phases of growth of a firm and the major concerns that accompany each phase. Let's quickly review:

Phase I - Small Startup firm	- Few assets and employees
	- Cannot tolerate any evaporation of assets
	- Major concern is survival

Phase II - Rapid Growth Firms	- Accumulate more assets and employees
	- Busy acquiring assets and capital
	- Rapid rate of growth of revenue stream hopefully outpace growth of expense structure
	- Major concern is fueling rapid growth rate and establishing company policies and systems.

Phase III - Return on Assets	- Rate of growth of assets and employees slows
	- Management shifts to prioritization and allocation of assets and employees
	- Focus is on proper depreciation and/or upgrading assets
	- Major concern is to maximize the firm's return on assets used

It is very important for your key people to understand that, as your firm progresses through to the next phase, you will experience rocks or obstacles to your growth. All firms experience these obstacles and the uneasiness that goes with them. The key is to see these troubles coming, and to line up our firms to take the safest route through the rapids of growth.

By discussing the meaning of each of these phases and what causes the problems or opportunities to appear, your people will sense a much greater feeling of having helped chart the course your firm will take through those rapids. This leads to a much greater sense of ownership for the quality of the ride the firm will take from both a financial and goal-fulfillment perspective. You will also find that people will be more understanding of the CEO having to shift positions within the raft.

Putting the firm's best players in the key roles is based on the needs of the company in each particular phase. A classic example of this is that of the firm's first outside sales rep (other than the CEO). He or she was hired during the early stages of growth to fuel the fire of rapid revenue streams. He or she is not asked by the firm to handle a lot of paperwork or to assist in the operational efficiencies of the firm. Just sell, baby sell! The company needs the best sales rep in the key role for the firm to survive.

As your firm's rate of growth starts to slow and you approach the rapids of the next phase of growth, where you start to focus on the return on assets your firm generates, you now expect the sales rep to start to live within an expense budget and to maintain a certain profit margin on their sales contribution. This tends to confuse the sales rep and irritate the new employees over on the operational side of the business.

A quick review of the phases of growth model should help all your people see why your firm needs the shift in job behavior. This is true for every position within your firm.

Once you have discussed the meaning of each of the phases of growth model, it is important to have each of your people plot where they think your firm is. Do not be surprised to find that different people in different departments within your firm have conflicting views of where your firm is located on the curve. That is exactly what you are trying to find out in a planning retreat. You want your people to be able to discuss their points of view to a point of understanding; not necessarily to agreement.

The raftmates will come to a much better understanding of what the entire team needs to do in order to be successful!

The next item to cover with your team is to have each team member present their SWOT analysis of the firm to the others. It is critical to have the scribe write down each of these points for future reference. Again, do not be surprised if you find some people feel a certain trait of your firm is a strength while others will see it as a weakness. The point of the exercise is to draw out the areas of inconsistency within your firm. Please take the time to allow questions and answers and to allow individuals to expand on their thinking for the group. Remember, you need to discuss the traits to the point of understanding, not necessarily to agreement.

Once you have discussed each person's SWOT analysis, it is time to do a Pareto model on them. Pareto, an Italian mathematician, was the first person to identify the 80/20 rule: 20 percent of your customers give you 80 percent of your revenue. In this case, we want to identify 20 percent of the strengths, weakness, opportunities and threats which impact 60 percent of the results of the firm. After grouping similar individual SWOT's, the team can assign a value to each and priority in accomplishing each element mentioned. Assigning each individual to champion their idea for improvement will draw natural hobby energy into the firm's ability to solve the weaknesses and threats that the team sees. You do not necessarily need to focus on the firm's strengths because they come naturally to your organization. Just make sure that an environmental shift does not render your strength useless. Does anyone care about how well Phillip Morris grows tobacco anymore?

The area of opportunity is always an intriguing set of traits to pursue with your people. Opportunities tends to start the creative people within your firm dreaming up new revenue sources for the future. Whole new market and service strategies come to mind. Hold onto those ideas because we will be using them shortly once we start building our pyramid of growth model for your firm.

Hopefully, you can see how in our first two exercises, plotting the phase of growth and SWOT analysis, that we have been opening up much needed internal communication between your key employees. We have prioritized the areas in which we feel we can have the largest benefit to the firm, and we have asked certain individuals to champion the post-retreat task teams that will carry them to fruition. We are well on our way toward becoming a stronger, more-cohesive team.

But, the bigger rapids are still in front of us, and we still need to help our people see what this firm stands for and how their individual contribution fits into the big picture. We will do that by filing a flight plan, and by creating our vision statement.

So far we have discussed points of differentiation between individuals or teams and how that slows our team as it achieves its goals. This difference in perspective is a naturally occurring phenomenon and is found in all firms to one degree or another. We will now need to shift your focus in the planning retreat toward building something that will help us achieve our desired goals and reap rewards for the firm and its employees.

A model of a pyramid of growth (page 69), has been adapted for use in our company from ideas by John Myrna in his book *How To Implement Total Quality Planning*. Each layer has a defined sequence and purpose in supporting the overall structure. You cannot work hard on action plans until your people know what your vision of the company is, and they know which products and/or services your firm will offer.

The pyramid symbol was chosen because it represents a manmade object, which has stood the test of time. The point of a pyramid is to generate favorable results: happy customers, company profitability, and fulfilled employees and CEOs. As the structure you build rises from its base, resting on your firm's mission or vision statement, the focus on each layer narrows or intensifies. Impacting either of a pyramid's sloping sides are the counter forces of hierarchy vs. modification and stability vs. flexibility which impact each of your employees to varying degrees. The actual process of planning and building your firm's pyramid model will foster teamwork; which builds trust. You should be able to achieve dramatically higher results, with far less stress, and your efforts will have a much longer-lasting impact.

Pyramid of Growth Model - From "How to Implement Total Quality Planning" by John W. Myrna, Myrna Associates, Inc., Leaders in Strategic Planning, 1-800-207-8192, http://www.myrna.com

The base of your pyramid will be in identifying the firm's values through a mission statement. Values are defined as a standard which drives your beliefs, and which you act upon. If your company already has a mission statement you need to make sure that your key employees know it and believe in it. They need to feel that they have input into what it says. It will not help you to have a canned mission statement that people do not value or believe in.

Many firms find that creating a visual picture of what your mission statement says, a vision statement, is far easier for your employees to remember. Most people in today's society are very visually oriented because of television, movies and bill boards. We are bombarded by ten of thousands of visual images portraying the results of using and or believing in someone's product or service. Visualize in your mind Joe Camel, the Marlboro Man or Ronald McDonald. We can all see them whether we believe in the product or not.

Your mission or vision statement should be a flight plan for your firm for the next three to five years. You may change a strategic goal from one year to the next, but your firm's values and beliefs should not waver with the economy, political trends, or social fads. Your mission or vision statement should clearly state who you are, what you do, and why you do it. There should be no question about what *drives* your behavior. Anyone looking at a good mission or vision statement should easily understand the various *characteristics that create your firm's* unique selling proposition or points of differentiation.

You are asking your people to commit to a set of company values and beliefs not unlike the way they commit to our country with the Pledge of Allegiance. With this company flight plan deeply imbedded in their belief system you can now be sure that your people will be making short-term decisions consistent with your firm's long-term goals.

The next layer on your Pyramid of Growth is to look at and develop your firm's product/martket strategy, your flow of commerce. This is the life blood of your firm and it must fit into your value and belief system, as outlined by your mission/vision statement. You should be looking at a one-to-three-year timeline, unless your industry has an even shorter life span than the software, and financial service industries do. For the product/service strategy you need to answer three important questions:

1. What product/service should we focus on considering external factors and competitive pressures?

2. How available are these products/services; that is, are they a generic commodity vs. a unique selling proposition?

3. How reliable are your sources of the product/service; that is, is there pending government regulation or artificial scarcity involved?

You need to be assured that the product or service from which you derive the majority of your revenue flow will be available to you from a reliable source, one that can withstand changes in external factors, such as weather, economic or social trends, speed of delivery, and that does not play into the hands of a much stronger competitor. I would not suggest a product strategy that focuses on beating Microsoft; uses tobacco, or is mined in Iraq.

The other half of the markets strategy is just as important. The four most important questions here are:

1. Who do we want to sell to?

2. Why do we want to sell it to them?

3. Where will we sell it from?

4. How will we get it to them?

You must first know what the product or service is, for that will determine your market strategy. You must be realistic about your strategy in this area. A good understanding of demograph-ics, geography, and cultural issues are essential. I would not sug-gest marketing a *Bisexual Guidebook For The Elderly* in the rural south; but it might work in college campus towns in the state of California. Three other issues you must attend to are:

1. Short-, mid- or long-term selling cycles and returns.

2. The impact of your strategy on available company assets.

3. Conflicts about flexibility vs. stability within your firm.

Starting up new products or services to be offered by your firm will be exciting to some employees and scary to others. You must also be sure that you can finance your new venture and that you have the proper equipment to deliver the finished goods to your customer. Items with long selling cycles (thus high selling costs) had better be associated with bigger ticket prices and higher margins. The same is true for perishable commodities or seasonal items. You should have a long list of prioritized opportunities from your SWOT analy-sis to help you get started with this strategy.

Once you have identified your mission or vision statement and se-lected an appropriate product/market strategy to pursue, you must now put together a proactive achievement plan. Setting strategic goals should be viewed within the context of a twelve-to-eighteen-month timeline. Your firm should have no more than five major strategic goals being pursued at any given time, so that the asset base and capital needs of the firm are not stretched to the limit. Each strategic goal needs to have an internal champion derived from the SWOT analysis to coordinate team members' activities, and to up-date the CEO, who will be the keeper of the plan of plans.

It is best to have a mix of passion and competence within the team-mates who will be working on the strategic goal. This synthesis of individual goals into team goals into the company's goal is what creates the power of this model. Three important questions to ask yourself about any strategic goal that you establish are:

1. Is it realistic?

2. Can you achieve it?

3. Will you win with it?

You can sometimes get caught up in the excitement of new opportunities from your SWOT analysis; that's one reason to have an outside facilitator and someone playing the role of Board of Directors during your retreat. If you can win—make a profit, and/or fulfill hobby energy—or not, is often a difficult issue to wrestle with!

A Strategic Goal Continuum (page 74) helps you to answer this question. In the left hand column you would write down your top five prioritized, strategic goals from your SWOT analysis. The continuum is a scale from 0 to 100 percent. Your job is to have the group select where they think your firm currently is on the performance continuum; with zero meaning you are terrible at the task and 100 percent meaning you are world class at the task. Again, your SWOT analysis will help you here. You need to be extremely honest with yourself on this ranking, and remember you are trying to see it as your customers would. You need to rank yourself from your customers' perception.

Let's say that you give yourself a score of 50 percent on the continuum of product availability. Most of the time you have what your customers need but sometimes you do not. Now, honestly score your competitors on the continuum, again trying to see it from your customers' perspective. If they score a 75 percent, then you need to work harder on this goal in order to win.

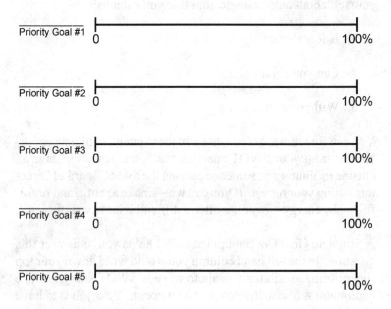

Priority Goal #1   0 ————————————— 100%

Priority Goal #2   0 ————————————— 100%

Priority Goal #3   0 ————————————— 100%

Priority Goal #4   0 ————————————— 100%

Priority Goal #5   0 ————————————— 100%

Strategic Goal Continuum

Remember that as you move toward the right on the continuum, towards 100 percent, it requires a lot more people, assets and time to show a smaller incremental improvement. The bad news is that it costs twice as much to move from 80 percent to 90 percent. But, your market may not give you a choice! The good news is that in your customers' perception, someone moving from 30 percent (relatively poor performance) to 60 percent, is getting a lot better than someone who moves from 80 percent (a very good performance) to 90 percent. Rank yourself and your competitors on each of these strategic goals. You will see where you can get your greatest return on time and energy invested.

Once you have decided on your five most-appropriate strategic goals, you now can develop specific, measurable, achievable, realistic, and timely (SMART) actions plans for each. This is putting the right things in the right order for everyone's benefit! They should be actions that will be accomplished in the next six-to-twelve months. Notice that each level on the pyramid has a shorter time frame. Let us define each letter:

S ·····The action must be SPECIFIC in nature, an actual number. You would say that a salesman must sell five jobs per month; not say they must get their fair share.

M ·····The action must be MEASURABLE in effect. You cannot say things like high quality or a lot or large amount. Instead, you need to define how the company will measure the action (i.e., score more than 90 percent on a customer satisfaction survey).

A ·····The action must be ACHIEVABLE IN NATURE. An action plan with great intentions but no chance for completion is not an action plan; but a dream instead. We get paid for results, not dreams!

R ·····The action must be REALISTIC to the environment. The individual or team must have the authority to accomplish their goal, along with the responsibility. You must also make sure that the individual or team has the proper knowledge, skill and aptitude to perform the task.

T ·····The action must be TIMELY in performance. It is important to put deadlines on tasks to give people a deadline. If you just said, "Sell five jobs," the sales rep could do that in one year by accident. By saying "Sell five jobs per month," the sales rep has a whole new attitude about hustle. You must also be careful to put completion deadlines for seasonal items and matching company cash-flow needs.

You want your action plans to be detailed enough to give the individual or team a glimpse of how their effort will look in real life, yet flexible enough that they can adapt to any rapid changes that confront them. That way, they know how to react in emergency situations, just like a rafting guide calling highside upon hitting a rock.

The "by who, by when" test is another good indicator of an effective action plan. This is especially true within a team task concept. The group needs to know who specifically is going to do each task so there is no confusion and no assumptions are being made. A good model to use for team task accomplishment is to have one leader, normally a motivator/influencer; one manager, normally a high level of job expertise and low level of need to be liked; and several technicians, people with strong knowledge and skills sets, but who may not like making decisions. If you are thinking that a SMART action plan could replace your firm's current job descriptions, you are on the right track. Job descriptions can be more than sheets of paper that describe what a job entails. SMART action plans, on the other hand, draw up a picture of what things will look like when you are done. And, they let everyone know who is going to do what, by when and how will we measure it.

SMART action plans are all about the timely completion of a specific task that is being measured. But on a daily or weekly basis, how would we know where we are? Once an employee finishes a certain task, how do they know what to do next, or how would they get ready for future tasks? That is where tactics and milestones come into play. The team or task leader needs to identify five to ten accomplishments which they all agree mean achievement of the action plan.

Tactics and milestones can be accomplished in hours, days, or weeks, depending on the task. A tactic is a repeatable way of performing a specific skill in a certain manner. It is how your company does a certain thing to maintain uniformity with the firm, even though many different individuals may be performing the task. Planting shrubs a certain distance from the house every time is a tactic of that firm.

Milestones are signs along the way that indicate how far we still have to go to the finish. They are a pacing device for runners and workers alike. A good manager can indicate what appropriate milestones should be passed, by a certain time, while the technician is performing the memorized tactic. Asking by what measure is a great way to insure alignment of the tactic being accomplished within the company values. Asking by what date will insure alignment with the company calendar and help keep people focused on doing an effective job. You must always watch for the effects of three seasonal influences on the by what date measurement tool.

1. Natural - weather related timing; i.e. swimming pools/ski industry.

2. Artificial - normal industry practices; i.e. new auto introductions, summer movie releases.

3. Calendar - societal norms; i.e. fireworks for July 4th and Christmas trees.

Another important function of tactics and milestones is their ability to help us react to rocks or obstacles in our business. With proper training we can improve our firm's ability to increase its synergistic reactivity. That is where we all react in unison to an emergency situation. If our firm has a swimmer, then everyone needs to not only do their job but also help recover the lost raftmate. This means that we need to cross train our employees so that they recognize their teammates' tactics and milestones, as well as their own. The ability to do this well is the sign of a true champion!

Summary

The final layer of the Pyramid of Growth Model is results. It is the pinnacle to our hard work and planning. If we have planned our work, and worked our plan, we should now be able to measure the results of our efforts. Results are the fruit of fulfillment in ourselves, our team and the company. When the proper effort is put forth toward a task that has stretched our abilities, we sense a great deal of fulfillment in those efforts. If, in addition to our individual effort, we have worked well within a team setting, the results can be even more fulfilling.

It is important to celebrate little successes; for they tend to lead to bigger ones. I will always remember the first time a raft I was in successfully ran a set of rapids. Our guide told us that all good rafters lift their paddles out of the water, high over their head, and tap them together in the center of the raft, and then slap them down on the river, as hard as possible, which creates a loud sound which can be heard above the noise of the rapid. As a team, we could not wait to run the next set of rapids so that we could celebrate again.

Rewards and what to do with them are key elements in successful management. *Fairly* sharing the rewards within a group, based on the individual effort put into the team results, is critical to building individual "buy in" towards the next effort. Splitting up the rewards equally among the individuals is not always perceived by them as being fair, and can then lead to reduced efforts by key individuals. The "buy in" is the key to the reward, for it leads to future performance.

One of the critical elements to increasing the value of a planning retreat is how you handle the "Moses syndrome." In the old testament book of Exodus, Moses goes up onto Mount Sinai to receive the Ten Commandments directly from God. The Hebrews receiving the Ten Commandments notice his radiant face and are afraid of him, instead of wanting to listen to him. This same effect, a radiant face, can be seen by your other employees on the faces of you and your key employees after you return from a successful planning retreat. You must plan carefully how you will discuss with the entire company, your vision statement, SWOT analysis, and action plans. Try to get all paperwork and documentation to the entire group within one week. Be sure to consult the company calendar for any potential conflicts prior to establishing any firm deadline dates. You want your first few action plans to lead to individual success, so that each person will buy into more action plans headed toward our company goals.

All of the action plans developed in step four of the pyramid of growth model need to be started within two weeks, so that the champions of those action plans can start with a sense of fulfillment. Speed of implementation is of the essence if we are to develop new "success habits" within our natural work groups in our firm.

The final step in a successful planning retreat involves the scribe rereading each key individual's retreat expectations. Review with the group how well you have met or exceeded the expectations. Reviewing what you have learned about each person's perspective on the firm's SWOT's will start you on the stage of a learning curve.

At the beginning of the curve, you are always "unconsciously incompetent." You are unaware you are not good at something! Then you become aware of your ignorance and you enter the stage of being "consciously incompetent." You are now very aware that you are no good at something. If you consciously decide to improve your knowledge (knowing what to do), skill (knowing how to do it) and aptitude (wanting to learn it) you have now moved into the stage of becoming "consciously competent." You now know you do not know everything you need to know about a particular topic.

To get better you must practice, practice, practice; until suddenly you can start to perform some minor tasks without even thinking about them. Only then, can you reach the highest level of learning by becoming "unconsciously competent." You can start to perform increasing difficult tasks without having to put much thought into it! Just by relying on your successful habits.

It is important to synthesize what you have learned with what you have skill and energy to accomplish. Creating a list of personal goals, with the appropriate action plans to conquer them, and establishing SMART milestones to measure your progress toward those goals, will lead you toward greater personal fulfillment. Combining those personal goals with company objectives starts you in the process of writing your job description. A group of individual job descriptions, now becomes the basis for developing a business plan, which should ultimately lead to company growth and profits and to a much greater sense of personal significance.

Long ago, we started on this journey by asking three common questions:

- Why isn't everyone on the same page in this firm?

- Why must we always appear to be "rowing upstream" against the flow of business?

- Why can't I find "good" workers who understand what needs to be done on a timely basis?

We have looked at a number of factors which have extenuated these problems:

- How technology is causing the speed and magnitude of change to accelerate in your firm.

- Why flexible, creative, and focused people are now our source of wealth in the Information Age.

- How different people map out their rapidly changing view of the world through their paradigms, and how we can learn to see their corner of the cube, if we just look for a different base.

- Why two people who are looking at the same situation see it differently because of their differing perceptions, and how learning to understand various personality and behavioral styles will help us see "their" side of the situation.

We have seen that in many ways running a business is like guiding a whitewater rafting trip, because businesses are very similar in nature to a river. We can certainly learn from this analogy how to place people properly within our firm to maximize our enjoyment and the safety of the trip. Also, by utilizing safety talks, rowing instructions, and emergency commands we can keep the number of swimmers in our firm to a minimum. Thus increasing our firm's synergistic reactivity!

Synergistic goal orienteering is a process that helps us develop a step-by-step plan that focuses your key peoples' "hobby energy" on mutually-agreed-upon prioritized, strategic goals. When your people are focused on a company objective, which has purpose for them individually, you should expect to see synergistic results.

Finally, we have looked at how to properly implement a strategic planning retreat for your firm. We discussed the proper timing, location, rules and roles your people should play in order to maximize your firm's return of the energies invested. Reviewing your people's retreat expectations, company SWOT's, and positioning on the growth model curve will indubitably bring out many inter-company issues which need to be aired. Remember, we are looking to create a point of understanding and not necessarily agreement with your key people on these points. Lastly, by building your firm's Pyramid of Growth Model, step by step, you will help your people see the light at the end of the tunnel. And, creating a mission or vision statement will help define your firm's values, beliefs, and identity for the next three years or more. Reviewing your firm's product or services offering strategy will help create both a short and long-term focus on the revenue flows into your business. Establishing SMART action plans, and putting them into the proper order, will dramatically improve your firm's ability to achieve its strategic goals. The fruit of personal fulfillment of achieving the desired results will lead to stronger teams and an increased chance of building future success. Remember, success is a journey, not a destination.

Stay Dry!

True Life Stories From The Real World

✦ *I Am In A Partnership That Works*

If opposites attract for the purpose of marriage, then opposites ought to attract for the purpose of partnerships. In fact, I believe that is one of the great secrets to partnerships and is an important ingredient to help them succeed.

Jim and Dave are definite opposites. If you were to meet them it would not take you long to see that fact. Jim is the quiet but calculative business person. He listens more than he talks but you know that what he is listening to is going into his calculative brain and will come out as a well-though-through idea. Dave is the exciting motivator that make things happen. He runs the field and most times runs a hundred miles an hour. When he sees a problem, he doesn't waste time in making dust fly and things happen. Jim is the salesperson who does whatever it takes to make you sign a contract. Dave will live by the contract but will confront you if you demand more than you are paying for in the original contract. Jim is someone who can research something from every direction and dot every I and cross every T. Dave is always looking at the big picture and is quick to move.

They started their company back in June of 1979. They were childhood acquaintances. Dave decided to bounce around a little and try this hand at Landscape Architecture and so he was attending school in the San Francisco area. Jim was attending college in the Los Angeles area where he was studying Economics and History. At the same time he was doing some very small residential landscape jobs.

Dave came home on a Christmas break, and as it usually happens, since he was studying Landscape architecture, he was asked to do a small landscape job for his mother. He called Jim to ask him where he could get the best buys on some plant materials. Jim told him and then put a seed thought in his mind that when he got back for summer break, maybe they could do some landscape projects together. In June they did their first job together where they were not only the company but the entire crew and from that point on, neither one of them would go back to school and thus began one of the finest landscape companies in Southern California.

For three years they struggled to run their business out of Jim's house and garage while they paid their dues in this business. In 1983 they moved into the big leagues. They landed a job that was $500,000 which required and allowed them to make a move to become a real business. For the first time they had an office outside of Jim's home and they hired people who could work in the office and help them run the business. They went from doing backyards to doing a $500,000 job. This awakened them to the fact that they could do and even liked doing commercial jobs while they continued to do residential jobs.

But within a few years they would find the next important level of change, doing custom residential jobs whose worth in dollars was something as large if not larger than some commercial projects. They found themselves doing large landscape projects for people like Dustin Hoffman, Michael Landon, Julie Andrews, Aaron Spelling and Ronald Reagan. They have worked very hard in developing the kind of reputation that continues to bring these types of residential jobs on a regular basis.

Then another major shift would come upon the company. They were selected as the contractor to perform a two-and-a-half year, $5,000,000 project. This brought another big spurt of new growth and new employees and new experiences. This new growth has brought on two new insights. First, they are more comfortable doing any kind of work but government work and second, they are now in a good place to evaluate whether they want to level off, grow some more or bring the company back down to the level it was before they got the big job.

During this time they took on something which has proven to be a mistake. They felt that since some of their contracts had an extended maintenance time to them, they could start a maintenance division. While this makes complete sense for some companies, it did not make good sense for them. They found that it not only became a money loser, but that it began to demand precious time that they did not have, and was beginning to deteriorate their construction sales.

I consider Jim and Dave to have an ideal partnership that is rich with its insights and workings. When I asked them some of the secrets to their partnership they would list some of the following items.

First and foremost, they have gained a mutual respect and trust for each other. Part of the reason this exists is because they were childhood friends. But this concept, once again, backs up the truth that partnerships can not be put together in a hurry. They demand that two or more people can take the time to get to know each other and to make absolutely positive that they can totally respect and trust the other person.

Any partnership where this kind of mutual respect and trust does not exist is certain to have problems. Too often, partnerships are formed for convenience or due to a lack of self confidence (see the story "I Had To Rid Myself Of A Partner") rather than two people coming together because they can help, because they have already formed mutual respect and trust for, each other.

Next, they are opposites and that keeps them from stepping on each other's toes, while it keeps them needing each other. These are two important concepts of a partnership. Too often if both partners are good at the same thing there is a power struggle going on as to who will lead in the those things. If both partners are good field generals, they will be having a power struggle as to who generals the most and the best. At the same time, if both are good field generals, who will look after the office and the administrative functions?

I again want to bring home how close business partnerships are to marriage. My wife and I are very much opposites. Where I am strong (finances, decision making, planning), she is weak. Where I am weak (sensitivity in decision making, spontaneity, stylishness), she is strong.

One of the keys to marriage and partnerships is that we cover each others weaknesses with each others strengths. If we have the same strengths and weaknesses how could we do that? I certainly do not want to be facetious with this statement but if my wife is just like me and vice versa, why do we need each other? It is the same in a partnership.

Another reason Jim & Dave feel their partnership has succeeded is that they don't socialize much during their off time. This would seem almost contrary to the way to bond and build a relationship of mutual respect and trust. However, there is an old proverb that says "Familiarity breeds contempt." There is definitely a line that must be drawn that says "We carry our partnership and our relationship this far, but no farther." Many a partnership has gotten into trouble, not at the office or in the heat of business troubles, but over a dinner party in which the partners and their wives or husbands got into some conversation that sunk the future of the company.

Another secret Jim and Dave have discovered is how they make the myriad of decisions that must be made. They have had to make numerous decisions for which they had absolutely no schooling or other preparations to help them make. They rely on two forms of decision making.

First, consensus. I have explained that form of decision making in the section "Taking Care of Business." Simply put, it means to come to a mutual decision through each party giving and taking from the original position so that the final decision is a mixture of both parties feelings and is something both can consent to. The other form they use only occurs when one of them has strong beliefs about something while the other partner has few or none. It is at those times that mutual respect and trust allows the one with the strong beliefs to carry the decision.

When asked about the things they were glad they did, they responded with two things. They were both glad they formed the partnership. They do not believe they would be personally, nor as a business, where they are today without their partnership. They feel they have forced each other to grow in many different dimensions.

They are also glad that they have a business philosophy that includes running a first class business that offers their clients exceptional service - something that everyone wants as their philosophy, but too few deliver as a reality. Their company succeeds at it because of their serious business philosophy, to have their employees feel like family because they care for them as people. Since people deliver a service for a company, they see these two philosophies as being interconnected.

But they are also equally dissapointed about two things that have impacted their business. At this point, they are at a place where they are tired and want to level the business out. The growth spurts of the past are not without their prices to pay and this is one of them. A more consistent and level growth pattern would have left them in a better position to maintain growth.

They also wish that they would have found a mentor or consultant to help them with their business years ago. They feel they have paid so many extra dues unnecessarily and could have avoided a lot of problems if they had found someone.

Partnerships like marriages, can go until "Death Do They Part," or they can go until "Divorce Do They Part." This partnership has tremendously favorable odds of going the distance because so much of it contains the right stuff.

✦ I Had A Strong Family Unit
That Made It Work

Dan is an extremely outgoing individual, one who takes pride in his accomplishments, but is quick to note that he is by no means perfect. He has been in the landscape business for almost twenty five years and contributes his success to what he termed "an extremely strong family unit."

Dan grew up in a family with five other brothers, all of whom were extremely close. They had their typical sibling "discussions," but could always count on each other when the times got tough.

Having a larger family of all boys, they were constantly called upon by the neighbors to do odd jobs, especially mowing lawns and the typical fall cleanup. At the time, Dan was only ten years old and enjoyed the satisfaction of turning a dull, overgrown patch of grass into a beautiful lawn. It was in the mid 1960's and the money they made was put away for college.

As the years went by the boys would start to take on larger jobs for the neighbors to include more detailed landscaping. In 1974 they started what they actually called a landscape company.

All six brothers worked for the company, and although the jobs were not large, they were beginning to build a sizable bank account.

Due to a shaky economy, Dan and his brothers were told by their father that if they wanted to attend college, they were going to have to make more money. The family discussed it and decided that the best approach would be for them to go all out with the company. To assist in this process, the brothers would take turn attending school. Half would go to school during the Fall and Winter, and the other half would go during the spring and summer.

So they formed a partnership between the six of them and began to build the business. Many lanterns of midnight oil were burned between the six of them. According to Dan it was almost an ideal situation. He feels that because it was family, they had an advantage that other companies didn't. Dan feels that they were able to have more challenging discussions than other companies, and if there was a problem late in the evening, they could discuss their thoughts right away without having to "wait until tomorrow." And because they were an extremely close family, they could say things that were personal and not have as much worry about hurting someone else or the individual becoming offended.

While attending his last two years of school, Dan decided that he would go to work as a construction engineer, and leave the business to his brothers. He was still helping them run the company but had let his involvement dwindle.

However, in 1980 Dan saw things happen that he felt uneasy about and decided to go back to his original love of landscaping. And as can sometimes be the case with the family company, things began to slip.

To this day, Dan can clearly remember the problems that were arising and the steps he felt he had to take to correct them. It still causes a little hurt when he discusses it, but as he said, "every cloud has a silver lining." Dan knew that problems were arising at an astonishing rate, and unless they were solved, the company would flounder. So on a night he dreaded, and called the "worst night" of his life, the family members got together and had a company meeting. It was at this meeting that Dan had to tell his brother that "there wasn't enough room for the two of them" within the company. Their management styles were different, and it was causing problems throughout the company.

And that was when something happened that took Dan completely by surprise - his brother agreed with him completely and dropped from the company! Dan couldn't believe what had happened. He had been so terrified of the scene it would cause, when there was going to be no scene at all! Dan's brother decided that what he really wanted to do was to form a paving company. So once again they held a family meeting, and it was decided that the landscape company would help Dan's brother form his paving company. Although they did not contribute large sums of money, they would rent their equipment to the paving company, and thus save the brother from having to buy any. The paving company soon grew, and according to Dan, is now making more money than the landscape business. Soon after the paving company was formed, another one of the brothers decided that it was what he also wanted to do, so he too dropped from the landscaping company and joined the paving company.

In 1980 the family "partnership" was dissolved and turned into the corporation which it is today. There were many reasons behind the change, but the biggest reason was to limit the liability. At that point Dan hired an Industrial Psychologist to analyze the company and try to get things settled for the long haul. He had known for a couple of years that there were problems, but did not know how to solve them. And soon he was presented with more information than he cared to know!

According to Dan, he soon learned that one of the biggest problems that he was faced with was employees. Sounds like your average landscape company right? Wrong. Dan's people were some of the best in town. The real problem was caused by the unhappiness of the employees. And all of the unhappiness was being caused because it was a family run business. Employees felt that they had no room to grow within the company because all of the top (best) positions were already filled by family members. The company was still small at this point and a large office staff was unnecessary.

So it was back to the family talking table to find the best solution - a solution that once again took Dan by surprise. Being a family of volunteer firemen for the local community, Dan's oldest brother decided that he was going to leave the company in pursuit of a career in medicine. And to really make things crazy, his other brother decided he was going to go after a different "calling" and is now working for General Electric. So now the "family" company was a corporation being run by two brothers. And things have been great ever since. Dan is very quick to point out however that at no time were there ever harsh feelings between brothers. They had just come to a point in their lives where they had to go different ways and pursue their own careers. Dan is also very quick to point out how instrumental his oldest brother was in starting the business and keeping it running.

Dan's business is doing extremely well today, due to what he called the "strong family unit." But he also feels that family owned business can be extremely dangerous. His reasoning is because most family businesses are run by Mom and Dad and then turned over to the children. He feels this causes problems because at this point, a matter of "who will run the company" comes into play. He believes that he had a distinctive advantage because his company was started by six brothers from the very beginning, and who would control what portion was predetermined.

Dan's company has had its high as well as low points -- just the same as every company. Dan states what he considers to be the low point is again, employees. But this time from a different angle. The office is now run by four people not associated in any way with the family, so that took care of the "growing" problem. But now he says that the biggest problem he faces is this: finding people who are both experienced and have the right attitude. He states that he constantly finds employees who have the right attitude, but know nothing about working in the field. Or on the other hand, can find extremely experienced workers, but their attitude is not one to fit in with the rest of the company. Dan attributes this problem mainly to the fact there is little formal schooling for landscapers. At first this may strike you as being totally incorrect, but is 100% true. How many of you reading this have attended a school where you were taken out into a field, given a shovel, pointed in a certain direction, and then told to dig holes to plant 25 gallon shrubs and 3 ten foot Pines? And then after you dug the holes were taught how to plant the shrubs and trees, wrap and stake the trees, then fine grade the field and lay 5,000 square feet of sod? Very few of you, correct?

All of you reading this know that most people learn the Landscape Industry from the school of hard-knocks. They get in the trenches, they work in the field, they watch as certain trees and plants die because they were planted wrong. There is little formal schooling, there are no magic number, no instant solutions. And Dan has come to face this reality as well. So instead of griping about the situation, he is developing his own training methods for employees, a method he says works quite well. He takes the employees to the field, shows them what has to be done, then shows them how to do it. He then tells them, "That is what I want done." and leaves. He does not hound them about doing things exactly as he does them. He constantly asks them for ideas on how to do things better, how to improve the working situation. And he says the results are incredible. He is rapidly learning that by leaving the employees alone, they produce the product better than he could!

Dan's high point during his career came when he bid two rather large jobs and was rewarded the contracts. His first thoughts were the same as all contractors, "OK, where did I goof, and how much money am I going to lose?" But when the jobs were finished, and the final job costing completed, he was in shock. Not only had they finished the jobs ahead of schedule, they had done them well under the anticipated cost and were going to receive a sizable profit on top!

Dan knew that the only thing that had made it all possible was the dedication of his crews and the long, hard hours they had put forth. So when the money was paid for the jobs, Dan decided to re-invest the money into the company. Some was paid to employees, but the rest was used to have a consulting firm come into his company and get it straightened out. He was shown by the firm how to streamline his operations, how to better manage things, how to make the company better as a whole. Dan feels that this turned out to be the best thing he had ever done. He also felt that it was extremely beneficial to the employees. He feels that his company is more productive than it has ever been, which in turn creates better attitudes in his employees, which make them produce a better product, which increases his bottom line.

Although mocked many times, Dan is also one who believes in sharing his experiences and knowledge of the industry with his competitors. Many people think he is crazy for doing this, and often tell him so! But Dan is not bothered by it all. So why does he do it? Why would he want to give away all of his trade secrets to the "enemy"? According to him the answer is quite simple - better competition. Dan feels that by helping other landscape companies to better themselves and teach them how to do things without "low balling," it makes them better competitors. And after all, that is what estimating a landscape job is - a competition!

✦ *I Am Managing A Company For An Absentee Owner*

In 1969, Chris was a 25 year old star on the New York stage. Those who were close to him wanted to groom him for Hollywood and the movies. Chris, however, was too idealistic and down-to-earth to be made into something he was not, so he left it all behind and moved to the Hawaiian Islands. He wanted to get back to working with the earth and farming.

For six months he worked for his father who was managing a vegetable farming venture for an absentee owner who had land on outer islands but a successful business in Honolulu. It was during the time working with his father when he met the absentee owner of the farm. That owner saw a lot of potential in the 28 year old "hippie" who loved working in the dirt and growing things. Soon the absentee owner was talking to Chris about starting some type of farming enterprise on a different 27 acres that he owned on yet another outer island.

Chris finds himself a very successful Landscape Contractor today primarily because of the many events that occurred during the next eight years of his career. They started growing organic vegetables which fit into Chris's outlook and lifestyle. The vegetables were an absolute success but the marketing and selling price of them was an absolute failure. Chris was putting in long hours tending to them and growing them the best he could, but he could not find a niche in the very competitive vegetable market. Nor could they bring themselves to invest the extra money he needed for the quality.

After four years of struggling with the growing of vegetables, Chris changed his farm product to growing ferns. This led him to the building of greenhouses in order to grow a variety of interior plants. The hours were still long and the money sparse but the excitement and satisfaction level for the "man who loved the earth" was still there.

But then there was another turn of events. The hurricane that hit the Hawaiian Islands in January 1980, leveled his greenhouses and damaged his plants. Rather than rebuild the greenhouses, he planted the damaged plants into a part of the 27 acres of the ground. It took them time to acclimate to the outside sun and rain, but they began to grow and become beautiful plants that could be used in the island landscape. Thus the nursery that exists to this day was born.

The next progression in Chris' career came through the help of an unselfish friend in the business who also had a nursery but was doing some contracting in the ever growing economy of the islands. This friend suggested that Chris had the resources and the talent to not just sell plants but to be a contractor and perform landscape jobs. Chris talked with the absentee owner whose faith in Chris had been well proven by this time. Together they formed the unique relationship that still exists today.

The nursery would still remain as it was, with Chris being the manager and paid accordingly. A new landscape contracting firm would be established with the absentee owner putting up the needed finances and shouldering the risk. Chris would get a salary for managing that company, as well as half the profits.

After ten years of operating that way, anyone who knows of the firm and of Chris's successes knows that this relationship has worked exceptionally well. What are the strong points that make it work?

Chris himself will admit that he is a rare breed indeed. Most people who are in the front lines of this business have an entrepreneurial spirit that causes them to want to do their own thing. Chris has a tremendous spirit but he is also very conservative.

That combination makes him fit this situation ideally. He loves the thrill of being in business but he loves the fact that he is not shouldering the risk. He gets the profit but could walk away if things turned sour.

Chris has become comfortable with the fact that, if he wanted to he could start a successful business but has chosen not to. He has come to know himself and has the luxury of not spending time and energy to find himself and thus he can spend that time and energy to make more money, which he has done.

Because he is not an owner, a part of Chris' profit may be taken out of the company. This has given him opportunity to do what I recommend all of you should do anyway. Have some form of other investments other than contracting if the profits must be left in the company.

Chris and his absentee owner have a tremendous mutual respect for each other. The eight tough years were a necessary ingredient in the building of that relationship. It was a time in which they got to see each other's true colors. The owner saw Chris' willingness and ability to stick it out and Chris had the opportunity to see the owner's willingness and faith in Chris since he kept supporting the relationship even when it was not profitable.

Chris' story goes to reinforce one of my beliefs. Contracting companies CAN be run by managers for absentee owners but the person who does must be a unique person. Most often, such relationships do not work because the persons involved do not fit the concept. Before any attempt is made to develop this type of relationship, the concept should be tested against the relationship described in this story.

✦ I Bought An Existing Business

Joe (not his real name), knew while he was in college that he wanted to own his own horticulture business, he just didn't know what kind. He educated himself in ornamental horticulture & standard business classes.

When he came out of college, he went to work for someone, so that he could get a real feel for what was going on in the real world. For four years he managed portions of a large nursery, knowing all the time he was capable and desirous of doing it on his own.

When the time came for the move he began to look through the "business for sale" ads in the trade magazines. He was not looking for a particular kind of business, but rather one that would fit a good buy that he could accomplish. He wanted one that had been a Ma & Pa operation where the couple wanted to retire. That kind and size of business would be affordable and ready for growth under his leadership. He also felt that in that type of business, Ma and Pa would finance the deal so that they could get their equity out more slowly and use it for retirement. Also, if Ma and Pa financed it, they would stick around a little while (but no longer that needed) in order to make sure things worked well and their investment was safe.

In June of 1984, Joe walked into the office as the new owner of such a landscape contractor company under the above conditions. In order to have the "right and workable" deal he looked for, he had spent eighteen months in looking and six months in putting the deal together. Patience not only got him the "right" deal, but has made the difference in why the deal worked. During that time he also attended seminars on buying businesses and read every book he could get his hands on.

Here are some facts on the deal. He bought all the equipment, property, and hard assets of the company at appraised worth. He paid some (as he looks back, probably too much) for the name and reputation. The owner financed eighty percent of the price over ten years. He is now into his sixth year of payments, and with recent appraisals he will be able to refinance the property, pay the original owners off, and reduce his payments greatly. The twenty percent down he obtained from a bank against the equipment that Ma and Pa took a second position on.

Joe is going to enjoy the classic method of making money through your real estate. He bought the business at a time when the area he was in was changing from small business to residential condominiums and apartments. Now he is surrounded by a residential area with developers willing to give him a sizable profit on his real estate just to build on it. He can then move out to the suburbs, or fringe of the city, and wait for it to grow around him again and sell that property for more profit. I have often said that I know landscape contractors who have made more money in real estate transactions such as I have just described, than they ever made in contracting.

Joe is now in this sixth year and has some insights into his purchase that he didn't have when he signed the papers.

First, he looks back at some of the right things that he did. He feels, that for him, buying an existing business was the right move. There are those who can bring up a business by the boot straps and enjoy the adventure. His personality lends itself more to managing an existing business better. He has the kind of personality that allowed him to come into a situation that someone else built and get the company to make the management transition from the old owner to him. Not everyone can do the kind of thing he could. He also points out that so many times he sees all the little things (permits, licenses, gas cans, small tools, etc.)

which are in place. He can't imagine all the work he would have had to do to get all these things in place.

He also credits his patience in finding the right deal and his willingness to read and study business purchases as a great help in doing a right deal. Too many people push too quickly to make a deal, any deal, figuring to work out the problems later. However, many times the hurried deal is so filled with unsolvable problems, that there is no solution to them later on. He cautions that since you probably will only do one of these deals once in your lifetime, make sure it is right.

Another important thing that he did right was to use the same bank as the previous owner used, at least initially. He felt that this was useful because the bank knew the business and its assets. It also knew of the seasonal nature of the business and its varying monthly profit picture. In fact, Joe says from day one the bank helped him put the deal together for his sake, and because of its concern for its long time retiring client, the previous owner.

However, Joe looks back to also tell you there are many things in the deal and in business that he would do differently. First, he made the mistake of using the company accountant and attorney. He figured, it worked with the bank, so it would also work with them. However, the accountant and attorney did not specialize in buy outs and thus, the deal cost more to put together and was not put together as could have been.

He also feels while he was patient in finding the deal, he became impatient when the deal started to come together. He negotiated for a longer loan at a lower interest rate but in exercising patience, he could have also got a lower price. Also, he would have spent more time and money on appraisals of assets and researching unsigned deals which he counted on and never became signed. Those things would have given him better leverage and substantiated facts in the final price negotiations.

He also made the mistake of putting all of his effort, planning, and energy into closing the deal and none into what he had to do in business after the deal was closed. This can be extremely deadly even though it worked out well for him. He wishes he had budgeted his overhead to include the payments on the business to see how much he would have to increase in business to make the payments.

You see, if a business with no payments to the previous owner is making a small profit of $40,000 at $75,000 in sales, possibly all the profit will go to the payments and in order to make any money at all, or have a margin of safety, the business must increase in sales considerably.

But Joe has also learned some typical lessons. After the deal he discovered he had bought a company with few systems and no computerization or anyone with computer skills. In order to grow and compete, he made systems and computerization one of his top priorities.

Also, he made the mistake of hiring a key person who was as green at the business as he was. This meant that neither one of them would intimidate the other with their superior knowledge but it also meant that they would have to pay the price on ignorance together.

He also paid the price everyone new to this business pays, believing that people are what they say they are. He has hired some key office people believing they could do certain things only to find out they could not.

He has also run into the proverbial bad, non-paying job. He had worked for a general contractor with great success on several projects. But on this project, the contractor became owner, developer and contractor for a HUD project. When the project started to run over budget, the contractor began to try to get the landscape contract cut in half. Those actions should have tipped Joe off to a problem and he could have pulled away clean before he put money into the project. The project went so over budget that HUD pulled their financing and now he is without payment for his work. He feels that when a general contractor ventures out into the unfamiliar territory of developing and owning a project, they bear careful watching.

But in retrospect, Joe has made a very good choice in both buying an existing business and in how he put the deal together.

The Trouble with People
and
Their Performance
and
What to Do About It!

Recognition for a Job Well Done

If you saw the first "City Slickers" movie, you'll remember when Curley, the tough old cowboy, said there was just one thing a person needed to know in life. Trouble is, he never told the dudes what that one thing was.

I, on the other hand, will not keep from you the one thing you need to know to build loyalty and esprit de corps in your staff. I want to tell you about one thing that will surely motivate your employees.

Several years ago, I had 12 of my clients gather in a hotel in Boston for a weekend of networking and brainstorming. They brought 22 of their middle-management employees along whom we asked to take a survey for which they could remain anonymous. We wanted to know what built loyalty in these employees and what motivated them. We gave them three choices and asked them to rate the choices from most important to least important by putting a number one by the most important, a number two by the next important, and a number three by the least important. The three choices were:

- *Money.* A raise or a bonus. Paychecks substantial enough to keep me from going to work for someone else for a larger paycheck.

- *A feeling of self-worth.* Knowing I make a difference in the world. A doctor or nurse or school teacher goes home some evenings knowing they've helped another human being. That experience is important to me, too.

- *Recognition for a job well done.* My boss should see me as a unique person with unique gifts and talents. When I use those unique gifts and talents on a job, someone should notice and recognize me for a job well done.

The Results: Do you know which of the three choices came in last on every survey but one—and then only hit number two? *Money.* Money is the least motivating reward and builds little loyalty. Do you know what was number one on EVERY survey? *Recognition for a job well done.*

Whenever I do a seminar for contractors and ask that question, everyone answers correctly as to how the survey turned out. That's because we all know in our hearts that the number one human need is to be recognized when we use our unique gifts and talents well. No one wants to be viewed as a hunk of meat bought with a paycheck.

So, if we all know that, why do so few of us ever recognize our employees? I believe that is because patting people on the back does not come naturally. We need to make ourselves do it.

Every year between January and April I get financial statements from clients from all over North America. There are two people who have sent them to me for over ten years. I wait for theirs with great anticipation. I cannot even leave the post office without opening them. Do you know why? Because during the boom times and the bust times, during a good economy or a recession, they have always made money. They are charismatic, motivating individuals, people who like people. And, they like to help people become the best that they can be. Because of that, they have never lost money.

One of those individuals has tried everything you can imagine to motivate his people and build loyalty to the company: bonuses, Hawaiian vacations, Caribbean cruises, turkeys at Thanksgiving. You name it, he has tried it. Several years ago he started the following program, which he says has done more to motivate his employees and loyalty to his company than anything else he has tried.

Look at the calendar in the back of this section (page 112). On the first Sunday is written "Family," on the second is written "Employee," on the third is written "Client," on the fourth is written "Supplier" and if there is a fifth Sunday during the month, on it is written "Friend." On Monday through Friday there is room to write in someone's name.

My colleague has several spies throughout the company. When they see someone doing something special, they write down that person's name and what they did. Every day, five days a week, fifty-two weeks of the year, minus holidays, the boss goes through those pieces of paper. He finds the person he feels has done the most exceptional thing. Then he gets in his car and drives to the job, finds that person, looks them in the eye, and thanks them for doing a good job. When he gets back to his office, he writes that person's name on the calendar along with a little bit about what they did.

Five days a week, fifty-two weeks of the year, minus holidays, he is doing the one thing that builds more loyalty and motivates people, recognizing his people for jobs well done.

And, recognition comes in writing as well as in person. The boss buys fifty-two note cards. He puts them in his desk near the calendar. During the first week where it says family, he writes a note on a card to a family member recognizing them for what they mean to him and his company.

On the second week where it says employee, sometime during the week he looks through all the names of people that he recognized the month before. He picks one that he feels was the most outstanding. He then writes a little note recognizing them for a job well done. He does not give the note to his secretary or put it on the employee's time card or in their pay envelope. He personally hand addresses the envelope and mails it to their home.

It's one thing to have one of your people going home and telling his or her spouse that the boss was on the job and recognized them for a job well done. The spouse might say, "Do you want your beer now or later? Are you going to eat with the family or sitting on the couch watching TV?" But, can you imagine what will happen if that news comes in the mail? While the employee is at work, their kids come home and read all about their parent. When his or her spouse comes home, they read all about their partner's efforts. When the employee gets home from work they read it, too, and they may receive pats on the back from family members. That kind of recognition will send their self esteem right through the roof!

On the third week a card goes to a client. On the fourth week it goes to a supplier. Maybe the boss once needed some material delivered to a job in record time. He called the sales representative for the supplier who only laughed at his request. However, that rep bends over backwards to get the supplies to the job on time. When the fourth week comes, the boss sends a card to that supplier recognizing that sales representative for a job well done. The owner of the supply house makes copies for every other sales representative and puts the card on the bulletin board. When the boss needs something else delivered again in record time the sales representative says, "It will be there. Are you going to send me another card?"

If there is a fifth week he sends one to a friend. I had an article written about me in a national magazine. He had read it. When that fifth week came, he took out a card and wrote a nice note telling me what I have meant to the construction industry and to his company, as one of my clients, in particular. I was gone when it arrived at my home. My kids came home and read all about their daddy, my wife read all about her spouse, and when I got home, I read it. It jacked me right through the roof!

Recognition comes verbally, in writing, and, finally, monetarily. That is why the people we surveyed put money last. Everyone saw it as important, but just a part of the bigger picture of recognition. Employers should put the recognition first, then add on the bonuses, the raises, the profit sharing. Next, how to hand out the money.

	Sunday	Monday	Tuesday	Wednesday	Thursday	Friday	Saturday
Family Name		Name Comments	Name Comments	Name Comments	Name Comments	Name Comments	
Client Name		Name Comments	Name Comments	Name Comments	Name Comments	Name Comments	
Employee Name		Name Comments	Name Comments	Name Comments	Name Comments	Name Comments	
Supplier Name		Name Comments	Name Comments	Name Comments	Name Comments	Name Comments	

"Good Job" Tracking Form

Bonus Systems:
The Good, The Bad, The Ugly

How do you reward with money in a fair and honest way? You must decide which of the many methods available will work best for your people, and what will truly motivate them to do their best work.

First, consider that your people should not be able to figure out any system you choose to employ. If a contractor makes it so the system can be calculated by everyone, then some employees will be doing their own calculations, which will be different than the boss—always in the employee's favor. This will lead to discussion and argument about each other's calculations, which leads to dissension, having the opposite effect from what was originally desired.

Also, one should be hesitant about any system that is too strongly tied into profit. If a contractor sets up the staff to receive a certain percentage of the profit based on their sales, then a contractor is taking on partners, in a certain sense. Because the employees' paychecks depend on profit, they will want to determine what are costs and what is overhead, because those items determine what is profit. The same holds true of any field person who is rewarded based on the profit of their individual job.

However, there are some systems that are based upon the profit the company makes. Employee Stock Option Program (ESOP) systems reward employees with stock in the company from money taken out of profit, which is then used to buy stock from the owner. A retirement program allows contributions out of profit to be made for the future retirement of all employees who are in the program. These programs are acceptable, but they have certain drawbacks of which a contractor should be aware before they implement them in their company. First, they reward everyone equally. Everyone who is in the program receives certain benefits, regardless of how much they may have contributed to the company's overall success. This can thwart individual effort to excel. Also, most of the retirement programs give out benefits based on how long a person is employed with a company. Therefore, the rewards are not based on performance alone, but on how long someone stays with a company.

Before I give you a bonus plan that I have found works best, let's look at another program that should not be instituted.

Some contractors reward field people by giving them a percentage of any money that they save on the estimated costs of the project that they are running. It seems like such a straight forward and beneficial idea, but let's look at what can happen.

An average field foremen has "beaten" the budget on a project and is given a bonus. The foreman is later in a bar buying everyone drinks when one of the contractor's best foremen comes in on the way home. The prized foreman asks the other where he got all the money. The celebrating foreman talks about the bonus.

Now, the prized foreman did not get a bonus because he was assigned to a job that is a problem in both estimate and site condition aspects. The prized foreman was put on the problem job to try to keep it up from becoming worse than it already was.

At the moment they are told about the other foreman's bonus, the prized foreman does not consider that he was put on the problem job because the contractor had faith in him. Instead he might think, "The owner doesn't like me and they like this foreman, so they give him the good jobs and me the bad ones." Or, they might think, "The estimator messed up on their estimate, and it is causing me to miss out on a bonus." Not long after, the contractor's prized foreman will come in, quit, and go to work for some other company. You can see that such a system—though it appears harmless—can cost a contractor a good man before the contractor wakes up and sees its real dangers.

Now, lets look at what will work. The system that follows rewards people for their individual efforts. However, employees cannot figure it out, or blame someone for being biased. A contractor forms a committee of three to five people who will be a part of evaluating the key, rewardable, employees, on a scale of one to ten (ten being the highest), for the following:

Customer satisfaction: Are the customers and architects happy with the quality of the work done, and with the cooperation given by your employee?

Do we get letters telling us how wonderful our foreman is on the job? Do the customers say that when they had questions, the foreman never acted bothered? The foreman would stop what he or she was doing and explain everything. The customers say they will recommend our company as long as we put that person on the job for their friends. If so, the foreman will rate high when it comes to customer satisfaction.

Or, do we get letters in which the customer says that at least the job got done but they never want to see our foreman again? Every time the customers had a question, the foreman acted bothered. The foreman never wore a shirt and used the customer's back yard as a rest room facility. This foreman will rate low on customer satisfaction.

Cost awareness: Notice the last word of the subtitle: awareness. That means beat the budget but it goes far beyond that concept. Do your employees beat the budgets that are "beatable" due to their personal effort? Or, on the other hand, do they allow good jobs to go bad, and to cost more than they should, even though they stayed within the budget?

Have you ever had someone working for you which thought your company's checkbook was a bottomless pit, and their role in life was to drain it? What was important to them was to get the job done the easiest, fastest way while expending the least amount of their precious energy. I have figured out why that happens. Such employees get the contract price stuck in their mind and they think all the money is yours. They don't realize how little is going to be left for the company, especially with them running the job.

We had someone like that with us for a little while. He was running a three million dollar job and he needed a backhoe on the job. He called on a Friday night and wanted one by Monday. I told him that ours were all being used so we rented one. He said he only needed it for three days. On the fourth day, I went out to the project. Do you know what was still on the job? The backhoe. I said, "I thought you only needed it for three days?"

He said, "I did."

I asked, "What is it still doing here then?"

He said, "I haven't had time to get it back to the rental center."

I asked, "Exactly when will you have time to get it back to the rental center?"

He said, "Maybe tomorrow." After I came unglued on the job, he never said a thing, but I could tell what he was thinking. He was thinking, "What is he bitching about? Why is he crawling all over my back? I am out on this project in the hot sun trying to get this job done. Get off my back." This is probably what he was thinking. No cost awareness.

On the other hand, I have had people working for me who were cheap, cheap, cheap. They were always thinking about saving money, our money. They would figure out a new way of doing something and how much it would cost and compare it to my costs in the original estimate and want to do it the cheapest way. And when they made a mistake which cost us money, they would come in and apologize for the mistake. They had their checkbook in their top pocket like they were going to pay us back for the mistake. We never accepted but the offer did make us feel good. Cost awareness.

Attitudes: Do they have a good attitude towards the company, their job, and those who are in authority over them? Do they help build up or destroy company morale?

I am talking of a specific attitude here even though there can be several more to consider. Have you ever had someone working for you who never made a mistake? At least if you ask them. But, they will tell you who made the mistake and it is never them. And, if they can't blame someone else, they will do something even worse: try to fix it themselves or cover it up. Later, you are served with legal papers. You are being sued. When you follow the lawsuit back, you find a problem or mistake this employee would not admit to when it happened. I have been in those kinds of situations and my company ended up settling for tens of thousands of dollars more than it would have cost us to fix the problem when it happened.

I want people who will run to someone when they make a mistake, get a solution, and learn from it. I have two daughters. When I took them out of the hospital and walked them the fifty feet from the door to the car and put them in their mother's arms, it was the first opportunity I had to talk with them as daddy to daughter. I looked into their cute little faces and told them that they were going to go places I did not want them to even get near. That they were going to do things that would make me mad. And, then I put them in their mother's arms. Do you know what the girls did? They went places I didn't want them to go and did things that made me mad. But, that didn't surprise me. However, we taught our girls when they had problems to come to us and we would help them solve the problems and learn from those problems. They did, too.

Next time you hire someone, teach them to come to you with problems, then—after they are out of earshot—say to yourself, "You are going to cost me money. You are going to screw up on me. You are going to make my life miserable." Then go back to your desk, knowing what will happen, and hoping the employee will run to you with problems to get solutions and to learn from the problem.

Equipment: Do they treat a contractor's equipment with respect, and see to it that it is not abused, and that it is serviced at the proper times?

I have had foremen who took the pickups home, washed and waxed them, and brought them back looking like they came off of a used car lot. Before they started the equipment they checked the fluids, and when the equipment needed servicing, they got it in to be serviced. When they operated the equipment, they operated it like they owned it.

Then I have had the opposite. We hired a superintendent to do a big job years ago. He was a budget beater. We were on the job for just two months and he had saved over $21,000 in labor. However, he was a one-man demolition derby. Every time he got on a three-cubic-yard loader, scratches appeared on cars in the parking lot. Whenever he got near a light pole, it fell over. He figured there was only one way to fill up the bucket of that loader, inertia. He could lay rubber with a loader and hit the pile so hard he would bounce up in the seat. Three months into the project we got rid of him. He was saving us money on labor but we were spending it to fix the scratches on cars, reinstall light poles, and fix the loaders. How does someone handle your equipment?

Paperwork: Do they turn in their necessary paper work on time, and is it filled properly? Do they turn in their time cards every day and keep up a job diary? Do they turn in delivery slips before or after they have been through the laundry twice?

I have figured something out about the people who go out into the heat and the mud and rain and build a project: They are Marlboro Men—and women. And, do you know what they think about people who sit in an air-conditioned office and do paper work? They think those kinds of people are sissies. So, when you ask them to do paperwork, they think, "Real men and women don't do paperwork, sissies do." Convince your field people to be an SFAM—a sissy for a moment—and do their paper work.

Quality: How long are the punch lists on their projects and how much work do you need to do over?

Safety: How many accidents happen on their jobs and what is happening to your insurance because of them?

After your committee has evaluated your employees, you'll need to make the following calculations.

Once someone in the committee has given an employee an evaluation for each of those items, the contractor adds up the numbers given for the seven figures—or the number the contractor decides to evaluate. The total is then divided by that number, to give an average evaluation for that employee by that committee member. The contractor then adds up the sum total of all those average numbers from all the committee members, and divides by the number of committee members. This, then, will give them that employee's average evaluation figure: Let's say 8.75.

The contractor then establishes how much reward money they are going to distribute, based on how much money the company has made in a given period of time. They then total the amount of all the rewardable employee's individual, final points. That total is divided into bonus money, giving a dollar factor per employee point. The contractor multiplies that factor by each employee's final evaluation number, and can then give his or her employees individual bonuses based on a fair, unbiased evaluation.

The following illustration gives you an example of a three person committee's calculations.

| | COMMITTEE MEMBER | | |
	#1	#2	#3
CUSTOMER SATISFACTION	8	7	9
COST PERFORMANCE	6	8	7
ATTITUDES	8	9	8
EQUIPMENT	9	10	10
PAPER WORK	8	9	10
TOTALS	39	43	44
AVERAGE PER MEMBER (DIVIDE TOTAL BY 5)	7.8	8.6	8.8

INDIVIDUALS AVERAGE RATING CALCULATION

7.8 + 8.6 + 8.8 = 25.2 DIVIDED BY 3 MEMBERS = 8.4 RATING

TOTAL RATING POINTS OF ALL EMPLOYEES

JOE	—	7.8
BILL	—	8.9
SAM	—	8.4
SUE	—	9.1
CHUCK	—	9.5
TOTAL		43.7

DOLLAR WORTH OF EACH RATING POINT CALCULATION

$18,000 IN BONUS MONEY DIVIDED BY 43.7 TOTAL RATING POINTS EQUALS $411.90 PER RATING POINT. EMPLOYEE LISTED ABOVE WOULD RECEIVE $411.90 X 8.4 (HIS RATING) WHICH EQUALS A BONUS OF #3,459.96.

Three Person Committee Calculation Example

Egg-Sucking Dogs

Recently, our company held a networking meeting for contractors in the mountains outside of Denver. For four days, the owners and key management people of nine companies gathered to discuss their problems and concerns, and to share ideas and solutions with one another. One of the subjects that came up was "the egg-sucking dog"; that is, the good employee who may have been with you for a long time—a key person who is making big money—who has adopted a bad attitude and has become a liability to your company.

At another seminar a person came running down the center aisle to me at the break.

"I'm from Tennessee," he said. "Do you know that there are such things as egg-sucking dogs? We had one down on the farm. A big golden retriever. He was friendly and loyal and loving. He would come up to you wagging his tail with egg yoke coming out of his mouth. He had been in the hen house rousting out the chickens and sucking down our egg money." That's what can happen in your company.

When it happens to you, and to someone in your company, it can be costly, not only to your company's finances, but also to your sanity. How do you keep it from happening and what do you do when it does happen?

First, as much as possible, do not hire friends or relatives. If you do, or if you already have, demand that you keep a professional, non-friendship relationship with them.

I find that too many contractors want to develop this friendly relationship with their employees. They want to have family

Bar-B-Q's together, and to go to outings together, and make their company just one big happy family. Early on, this type of approach has its benefits. But as time goes on, it will usually come back to haunt you. Not because people are purposely going to take advantage of your friendship, but because somewhere down the line, they may begin to expect special favors because of their special relationship with you. When those things do not happen, their attitude will begin to change, and you will find yourself in a very tough position. Or this old friend may have old resentment toward you that will begin to affect his behavior—and your company.

I know someone who was the quarterback of the football team in high school, and he had a friend, the friend of a lifetime, who was the halfback. They dated two cheer leaders who were also friends of a lifetime. When they graduated from high school, they later married their high school sweethearts and they all continued to be friends.

A year and a half after graduation, the quarterback's dad died. He had owned a small construction company, which the quarterback took over. He hired his halfback friend. Together over eighteen years they built the business into a $25,000,000-a-year construction company. Then, after eighteen years, his executive vice-president and friend of a lifetime became an egg-sucking dog.

He was good. He could bid jobs, troubleshoot jobs, negotiate jobs, negotiate change orders and even read financial statements. He was loyal to the company and his friend. But he had become an egg-sucking dog.

Later, the employee told us what had gone wrong. "When we were in high school and we won a football game, the quarterback was in the middle of the field surrounded by people who were giving him all the accolades. He didn't have a grass stain, mud stain or blood stain on his uniform. He had never even hit the turf the whole game. I stood on the sideline with very few people saying anything to me, but my uniform was covered with mud stains, grass stains and my own blood stains. He just gave me the ball and I hit the line. When someone was gang tackled, I was gang tackled and when someone stepped on a hand with their cleated shoes, it was my hand that they stepped on. Yet, he got all the accolades. And now, he is the president of the company and I am just his vice-president. I have been living in his shadow all my life."

Those two men are still friends of a lifetime. The president of the company found a way to confront him and let him go. The egg-sucking dog now has a $10,000,000 a year construction company and my client is still running his $25,000,000-a-year construction company. Since this happened, the egg-sucking dog has said to me, "I was too loyal, but did not have the guts to quit. I am glad my friend had enough guts to let me go." Sometimes a good, old employee needs to get a new start somewhere else. And, as for the owner of the company, as Kenny Rogers sings, "You've got to know when to hold them, and know when to fold them."

Now, I will admit there are exceptions to this rule, but they are just that, exceptions. The rule still remains that you should not develop overly friendly relationships with your employees.

One of the reasons why many people get too friendly with employees is that they find that it is lonely at the top. This is why I feel that every owner needs to be a part of some organization of peers in which he can share his or her problems openly with other who are in the same position. This way, they can avoid this kind of relationship with their employees.

Another thing to do is to look for early warning signs that a good employee is becoming an "egg-sucking dog" and then take action while he or she can still be turned around. One of the great regrets expressed by those who have faced this problem, is that they did not take action soon enough. They saw the change slowly taking place but did not want to acknowledge it, or deal with it at the time. They kept telling themselves it was a temporary problem that would fix itself, and they kept blaming it on something the employee was going through. They convinced themselves it would change very soon when the employee passed through the bad time. Many of those who did not take action until it was too late felt that if they would have done something sooner, there would have been a much better chance to save that employee. We walk a very fine line between overreaction and procrastination in these cases, but with very few exceptions, it is better to confront the situation sooner than later.

However, I must point out this sad fact: Most people with whom I have talked feel that, in the majority of cases, the only solution for good employees who have gone bad, is to fire them. Everyone agrees that every reasonable attempt should be made to salvage them. But eventually you must be enough of a realist to know that, in most cases, those efforts will fail. You must decide on how far you can afford to go with this person, setting a limit on your time and efforts to salvage them, and then when those are not met, be at peace in letting them go, knowing that you have done all you could.

Do not internalize the guilt. My client, the quarterback, kept asking himself what he had done to his friend of a lifetime. Well, he was not perfect in all of his dealings with him, but he had not done anything that bad. His friend was paid a six-figure salary, had a company car, bonuses, and stock options. How can you make up for a friend's feelings that go way back to high school?

That fact leads me to one final thing that can be done to keep a good employee from going bad. I feel that sometimes this problem occurs because the employee loses one of his most important motivators, that is the need to impress someone. He has worked for you, has done his best, and now feels that he has nothing else to prove. Your job is to keep such people in positions where they feel that they must prove themselves every day. This can be done by giving them more difficult jobs, working with owners and architects who are more difficult to please, or by giving them more responsibility. However, this may be something that just keeps them for as long as possible. There may very well come a time when they think that they are better than you, and that you cannot do without them, and that there is nothing left for them to prove. At that point do yourself—and them—a favor. Send them down the road to another company where they must start proving themselves to someone else all over again.

I cannot say it enough: this is a people business. The most successful people in this business are savvy about the dynamics of human work relationships.

Four Personalities
That Must Exist in a Company

The management of a construction company and individual projects requires a team effort. The team must be made up of four distinct personalities. Each of us has only one or two of the four personality types within us. If you are a small company, you must do the things that all four personality types must do. That is why you often get frustrated and may not do all the things you need to do to run your company successfully.

Below are the four personality types related to the building of one project. As you read this, ask yourself this question, which of the two am I? It is important for you to know so that you can surround yourself with people who are diametrically opposed from you in personality.

The first type I call the "Idea Person." The Idea Person may be an architect, interior designer, computer programmer, or an artsy person. You can always tell an Idea Person because he or she will say, "I have an idea." Most ideas are stupid and impractical. But there are the few good ones, and every team and every project needs an Idea Person to put the company on the cutting edge with those few good ideas so the company can make the most money.

I have been involved with a company that was being run by the third generation. The grandfather started the company in 1908. They have his original truck and loader restored and parked in the yard. They say to me, "We are still doing things like our granddaddy did them."

I say, "I know, I can smell it. It smells like formaldehyde or mothballs around here."

The people in this company have not had a new idea in 30 years. They are so far from the cutting edge that their company prospects are as dull as a butter knife.

I have 20 acres and I am going to build a brand new hotel and convention complex. Who is the first person I call? The architect. He comes out on my 20 acres and says, "I have an idea. Let's build a hotel that looks like a giant escargot."

I say, "That's a good idea, draw it up." Every company and project needs an Idea Person in order to keep the company current and profitable.

The second personality type I call the "Happening Person." Happening people are contractors. They like to take someone else's ideas and create them. Where there is nothing they want to see something created by the force of their will—and at their command. They are focused and driven, and are not satisfied until they have made something happen.

I have 20 acres and I am going to build a brand new hotel and convention complex. Who is the first person I call? The architect. He comes out and says, "I have an idea. Let's build a hotel that looks like a giant escargot." He draws it up and I put the job out to bid. I receive three bids and award it to a contractor. The contractor calls in the surveyors, the bulldozers, the concrete people, the steel erection people, the glazing people, the roofers, carpenters, plumbers, mechanical people, the electricians, the pavers, the painters, the carpet layers, and the landscapers. They make that hotel happen. Why? Because they are happening people.

Every company and every project needs happening people. Have you ever been to a meeting where everybody has ideas? And, after everybody has told their ideas you leave that meeting and you have a funny feeling, which is that nothing is going to come of all those ideas and nothing is going to change because of those ideas. Do you know why? Because there is no Happening Person there who will go through those ideas, take one and be focused and driven until they have made that idea happen and see it working.

The third personality type is what I call the "Managing Person." Managing people are coaches of baseball teams, hockey teams, basketball teams and football teams. Behind the white line of every football field stands a head coach. He has hired an offensive coordinator, a defensive coordinator, a line coach, a backfield coach and a defensive backfield coach. They are all managing people. They don't go out on the field and take a snap, they do not run the ball, pass the ball, catch the ball, block any one or tackle anyone. They manage all the details of what goes on in the field. They make sure the team is in the right city with the right uniforms at the right time. When it is time to kickoff or receive, they make sure the right number of players is on the field. When it is fourth down with a yard to go, they decide if the team should punt, go for a field goal or try for the one yard.

I have 20 acres and I am going to build a hotel and convention complex. Who is the first person I call? The architect. He comes out and says, "I have an idea. Let's build a hotel that looks like a giant escargot." He draws it up and I put it out to bid. I award it to a contractor who makes that hotel happen. Then they give me the keys. Now who do I need? I need a hotel manager who will hire an assistant manager, a manager of guest services, a food service manager, a housekeeping manager and a catering manager. They see to all the details of what goes on in a hotel. They make sure that you get checked in, that the rooms are clean and have the little shampoos that you can steal. They make sure meetings are set up and food is prepared.

Every company and every project needs managing people. They make sure the payables are paid, the receivables received, payroll is made, materials are ordered, paper work is done, and work is bid.

The fourth personality is the Maintenance Person. Over 70 percent of the American work force are maintenance people. You can always tell a Maintenance Person, because they will always come to you and ask what people they should take, what tools they should take, what truck they should take. They want you to tell them exactly how to do the job and they will do it just the way you tell them.

I have 20 acres and I am going to build a hotel and convention center. Who is the first person I call? The architect. He comes out and says, "I have an idea. Let's build a hotel that looks like a giant escargot." He draws it up, I put it out to bid and award it to a contractor. They make the hotel happen. They give me the keys and I hire a manager who hires other managers and hires 200 maintenance people. People who come in every day and ask what they are supposed to do, what rooms they are to clean, and what food they are to prepare, and how to set up each meeting room. Every company needs maintenance people who will get the work done. They are the ones who are really building America and they are the engine of American business.

Do you know which of these personalities come easily to you? It is important because you need to surround yourself with people who are different from you, who can be strong in areas where you are weak. If you are an Idea Person and a Happening Person, who is going to manage and maintain the ideas that you come up with and make happen? If you are a Managing and Maintenance Person, who is going to have the new ideas that make money; who's going to make them happen in your company and on your projects?

So, build yourself a good—and a complete—team by including all these different types of people in your company. Now, let's look at what effect one large group of people has on your industry.

The Baby Boom Phenomenon

As I travel around the country I become more convinced that this country and our industry are being greatly affected by "The Baby Boomer Phenomenon"; the influence of what is now the largest demographic group, people born between 1950 and 1970.

If time is a python, the Baby Boomers are a pig being digested by that python. Their impact and influence are that noticeable. All of the advertising, music, and cultural styles cater to the demands, wants, and needs of them, because they are the most dominant money spenders around. And, their effect is anticipated to continue for 20 years or more. I fully expect that when I am in my sixties I will be watching a TV commercial in which a gray-haired man gets out of a golf cart drinking a soft drink, and saying, "Uh-Huh." They will still be trying to capture my dollars—for I am a boomer, too.

Now, let me tell you why and how this boomer generation is impacting you and your business. First, let's talk about your labor and management people.

The Loyalists

I read an article several months ago that really impacted me and really turned on a light by explaining what I have been seeing and hearing. It depicted the boomers as "disloyalists" and their fathers as "loyalists." It said that our fathers went off to WW II and the Korean conflict and came back very patriotic, very American, and very loyal. They went to work for a company and intended to retire from that company. Any moves up in their careers were within that company.

My father went to work for a company in Denver and retired from there after 32 years. Never once did I see him looking in the help wanted ads of the paper. Never once did he consider quitting his job to advance in another company. He would always say, "I have a job." If another company had offered him a job for more money, he would have turned it down because he was a "company man."

The Disloyalists

The boomers—the disloyalists—saw that kind of loyalty rewarded with the shut down of the steel industry in this country, the closing of factories, and massive layoffs. They saw their fathers laid off at 55 or 60 and sent into a job market where they were forced to take menial jobs for which they were overqualified, or they saw them forced into early retirement.

Seeing what happened to their fathers built an attitude of disloyalty into the boomers. Their attitude has become, "Ask not what you can do for your company, but ask what your company can do for you." The average boomer is with a company for three years before he begins to look around for another job. He or she has usually moved on within six years.

Think about the boomers' fathers and the boomers you have hired. Isn't it true?

Many companies I work with are frustrated with the turnover in people they have. They spend money and time to train people only to have them leave. They ask, "What's wrong with my company and me?" Many times, it's simply the fact that they are dealing with the general disloyalty of the boomers.

I have heard of an idea of how to handle this national problem. Some companies are making it a policy that if people want to leave, they are asked to let the management know of their desire.

The company will help them find a new position. Management then has two important opportunities. First, the company can help the employee get into another situation and company where they are not in direct competition with the first company. Second, management also has more than a two-week notice during which to replace the person and make the transition as easy as possible.

Next time a disloyalist resigns from your company, realize the problem may not be you or your company, but a characteristic of the boomer attitude toward work.

Building to Satisfy the Pig

I see another way that the boomers are impacting our business. When we are in recession in the construction industry, many want to blame the president, the Congress, the S&L Crisis, interest rates or high capital gains taxes. All of these things are contributors, but lets look at another big contributor.

The boomers are said to be about 90 million strong. Their parents bred for themselves an economic boom. I am at the early end of the group, and when my friends and I got married, there were not enough houses, cars, goods and factories made for the boomers, or stores to sell things to them. We boomers put an almost instant demand on the market place for all these things. This brought about the great building boom in the late '60s and '70s. This country was building to satisfy the pig in the python.

But, we boomers were not as smart as our parents. In the '60s and '70s the talk was of population explosions and the need to replace yourself with only two (or fewer) children. I have two children. All my friends have two kids. We boomers have bred for ourselves an economic recession. Behind the pig going through the python there is a mouse when there should have been another pig, or even a cow. I see the construction industry facing a long time before another building boom, because the industry built for the pig in the python and what has been built will satisfy demand for many years to come.

My daughter is getting married in a few months. No one needs to build her and her husband a house, or factories to make goods, or stores to sell them. There are more than enough of all these things.

The Boomers and Service

There is one more thing we need to understand about the Boomer Phenomenon. Boomers are splitting into two groups. The "Rich Boomers" and the "Poor Boomers." The Rich Boomers were born between 1950 and 1960. They got into their houses when you could afford them and into careers when jobs were plentiful. The Rich Boomers inflated the prices of goods and got a head start on their careers, leaving the Poor Boomers—born between 1960 and 1970—in their wake.

The Rich Boomers have built their houses and are in them. They are not going to be out building things, but they will have the most money to spend of any age group in the country. What will they spend all the money on? Service. They will not want to mow their lawns, trim their shrubs or pull their weeds. They will not want to plant a few trees and shrubs to upgrade their landscape. Consequently, one of the fastest growing segments of this business is and will be maintenance. All across the country I have heard and seen the same things; contractors talking about how maintenance has paid the bills, how it is the only part of their business that is growing. Companies who never even considered maintenance three years ago are now scrambling to get into it. Do not doubt that maintenance billing in this nation will exceed the billing for new construction someday.

The Boomers and TQM

When people came back from World War II and the Korean War, they started companies here in the United States. And, they needed a management style to run those companies. The major point of reference of running a company was military.

Consequently, they organized their companies after the same style of management that they knew best—the military. There was a general (the owner) and then various other officers (the administrative staff) and, finally, the sergeants and privates (the field people). During this time, the originators of Total Quality Management (TQM) tried to get that style of management started in the United States. This has become a buzzword in the construction industry.

Without going into a long description of TQM, we want to give you a basic understanding of its overall concept. TQM in its simplest definition is the demilitarization of the management style that exists now in the United States. It is the building of a management style that includes all people from all facets of a business. These teams are facilitated by a person who includes all participants in the management of the company and the solving of problems.

However, because most people were schooled in a military form of management, TQM was not accepted in this country. Because of this, the originators took it to Japan where, because of their bad military experience, it was accepted and instituted. Now it is coming from Japan back to the United States. People are recognizing that it is a major part of the reason Japan has become the major world economic force that it is.

With the boomers a cultural revolution started in the United States. Many say its impact, as seen in overall history, will be as great as the industrial revolution. That cultural revolution brought about a new generation of thinkers and people who desire intense involvement in whatever they do in life. It is that attitude that has made TQM possible and so successful in the United States. Every business book I have read in the last five years speaks in one way or another of how the cultural revolution is causing American business to rethink, retool and restructure the management of their companies.

The Boomer Phenomenon -- it certainly has, and will, continue to affect our business.

Attitudes to Guard Against

This is, by far, one of my favorite subjects! It is probably the most over looked principle in construction today. Remember, construction is a people business and there is nothing more important when a contractor is dealing with people than attitude. The attitudes of a contractor's employees, especially their estimators, can cost thousands or make thousands. Let's consider a few.

The first has to do with an attitude of contention that can exist between the field and the office. A person gets promoted to being an estimator, so he starts wearing a suit and his hands are always clean. He drives out in a fancy car with the air conditioning on and the stereo blaring. He gets out and looks over the job with his nose in the air.

"Don't get dirt on me," he yells, as he jumps away from some construction task being performed. He walks up to shake hands with the superintendent, but refuse to do so, because the superintendent's hands are dirty. Then after looking around piously, he leaves the project with the air conditioner blowing and the stereo blaring.

Do you know what they say about that person after he leaves the project?

"We're going to have his job." And, do you know how field people are going to try to get that job? By letting the costs the estimator bid overrun in the field. Do you know who's going to be losing thousands? The owner of that company.

"Oh," you say, "The owner will fire field people who do that." Yes he will, until he figures out that the estimator is the problem. Then the estimator will be fired.

Contention between the estimator and the field costs multiple thousands of dollars to the construction industry every year. I have actually walked into contractors' offices and heard the following kinds of conversations between estimators and the field.

"I know it can be done in _____ amount of hours. If you can't do it for that you ought to be fired," says the estimator.

"Well," says the superintendent, "If you think it can be done for that, get out here and do it for that!"

And, then they both slam down their phones on each other. Do you have any idea how much that costs a contractor? Thousands upon thousands of dollars!

I have always liked field people. I treat them with utmost respect and courtesy and always show myself ready to learn from them. They have made me what I am today. They have made me look good when my estimates have looked bad. I've had field people come into my office after they had heard or seen a mistake in my estimating and say, "Don't worry, we will cover it up. We'll get it done for what you figured." And, they do. No one has a cash register big enough to total what that kind of attitudes can make a contractor.

Sometimes people take this attitude, "We've got to have this job!" Work is slow. The contractor has equipment sitting in the yard that is not working. They also have two superintendents on the same job and they could sure use another job for one of the superintendents. So they look at this job they are estimating through eyes that say, "We've *got* to have this job." With that attitude they begin to estimate the job. Do you know what they are going to do when they put that estimate together? They are going to say, "We'll do this or that cheaper than we normally do it." Not because they will, but because they've *got* to have this job. They will convince themselves that certain methods or materials will work because they will make them work—because they've *got* to have this job. When they're all done with the job, it will cost them what it will cost them, but they will not have enough money in their estimate after having lied to themselves in order to have the job. But, costs are costs are costs, and they have missed the important concept of what estimating is: the science of arriving at what a job *costs* their company. When they lie to themselves that way they have no idea where they lied or how much they lied about. If they want to go after a job, know what their costs are, and then trim their overhead and ask for their lowest profit, but don't lie about costs. This attitude costs the company money in the end.

Another attitude to look out for is an attitude that says, "This is an award winner!" If we get this job, we'll get a plaque at the national convention in January." So a contractor bombs the job by $20,000 or $200,000 to get their award. Put that money in the job and send it to me. I'll buy you a whole room of plaques.

A similar attitude goes like this, "If we get this job, it will lead to a lot more work. We can put our construction trailer right by the highway. On its side we will put our name, Joe Blow and Sons and Daughters Construction Co., Inc., Ltd. People will drive by and see our trailer and run right into their office and call us to do their jobs." So they lie to themselves about the job and bomb it in order to get more work. Except the other work doesn't come or, if it does, people expect it at the same cheap price. Bid every job like it's going to be the *only* job a contractor ever will do for anybody.

Developers love to use this advantage. They approach a contractor with a set of model homes and tell them, "Do you see that apartment project over there? If you do my models real cheap, you'll be in on the ground floor to do these apartments." And, while that contractor is going broke on the models, the developer is over talking to another contractor about the apartments. "Do you see those shopping center plans over there?" they ask. "If you do these apartments real cheap, you'll have the inside track on that shopping center." And, while that contractor is going broke on the apartments, they are talking to someone else about the shopping center. On and on the circle goes, because there always seems to be someone who will grab at the ever-elusive carrot. Bid every job like it's the only job you will ever do for that person because it will probably will be.

Another attitude to look out for: "We'll bid this job and make money on the change orders." Who says a contractor will get the change orders they need at the price they want?

Change orders can be very risky. That doesn't mean a contractor does not apply for them, but don't even go after a job because of them. In fact, do you know what most hassles between owners and contractors boil down to the final analysis? Change orders! One of two things usually happens when it comes to change orders. First, the contractor will not take into consideration all the costs involved in the change and thus cheat themselves. Or, second, if they include it at all, the owners refuse to pay the amount of their change order.

Some final attitudes to consider are "optimistic versus pessimistic" attitudes. Some estimators are real optimists. They see every job estimate through rose-colored glasses. The job will have no problems. They don't figure labor to pour and finish the concrete, because the concrete will just fall in place. They calculate all labor at optimum performance and production rates. This job will go so well—as they see it—that the architect or owner will give the contractor a bonus check when it is all done. When the estimator comes for a job review, the contractor looks to see what they left out of the estimate.

Then there is the pessimist. They come to work looking as if the dog died and the cat ate it. Whenever they look at a job, they can tell you everything that can possibly go wrong on that job. When they estimate the job, they take all possible problems as fact and include money to cover their costs. When they come in for a job review, the contractor looks for all the extra things they have put into the estimate.

"Well," you say, "I would get rid of people like that right away!" Why? They are you and me, and we are they. Each one of us, every day of our lives, fluctuates between optimist and pessimist.

It's Monday morning and you are on your way to the office. It's 7:30 A.M., yet you're the only car on the freeway. Your windows are down and the temperature is a perfect 72 degrees. Your favorite song is on the radio and a bird flies alongside your car chirping the melody of the song. You arrive at your office, and the first thing you smell is freshly brewed coffee. You go by the secretary's desk and notice that over the weekend every expected accounts-receivable check has come in the mail. Everybody that owes money has paid. You go to your office, and there in the middle of your desk is a brand new set of plans and specifications reeking with ammonia.

You open them up and say, "It's great to be alive and to be a contractor in my city." You begin to estimate that job feeling enthused about contracting and your company, because at that moment you are an optimist.

Then the phone rings. It's one of your project superintendents. They started up one of your pieces of equipment and left it in neutral. It rolled down a hill and ran over a little old lady and then crashed. The equipment is totaled out and the lady is in the hospital. You hang up the phone and go back to work. Twenty minutes later the phone rings again. It's your insurance company and he's calling to inform you that your insurance policy has just lapsed on that piece of equipment. You go back to your bid a different person. Now you have become a pessimist.

"Oh no," you say to yourself. "This job has got a huge hill on it. We're going to lose two pieces of equipment over that hill. And look here's an old folks' home right by the project. We'll send three old people to the hospital." And, so you take your estimate sheets and add three brand new pieces of equipment. One, to replace the one you just lost, and two, to replace the ones you're going to lose. You also add in hospital costs for four people, the one who's in the hospital and three that you probably will send to the hospital.

What's happened? You've gone from an optimist to a pessimist. That is human nature. But, you had better take that fact into consideration because your attitude, and how you feel about a job, *will* influence your estimate.

When an estimator goes in for a job review, he should be brutally honest and transparent. He should say, "I really like this job. It would be fun to do." Then others can look at the estimate and say, "OK, what did he leave out?" Or, he should say, "This job stinks. If you want to bid it, that's your problem." Then they can ask "OK, what extras did he throw into this job?"

Remember, estimating is a *process*. It begins with the normal production rates, then an estimator massages the labor based on the exceptions. Then he takes his estimate to someone else and convinces him that it can be done for what he figured. And, finally, he takes into consideration attitudes and how they have affected his estimate.

Building Good Communication
In Your Company

There should be no doubt in any contractor's mind that he is in one of the biggest "people businesses" in the world. He should also have no doubt that one of the most important things in working with people is the ability to communicate. Here are some practical ways you can develop communication skills in yourself and in your staff.

Before working on individual skills you need to sit down with your people and sell them on how important good communication is to your company. Tell them about the things you'd like to do to improve communication. Make a plan, with their help, so that everyone knows what you are going to be doing to improve communication. Let them know that some of the things you will be doing may come across as very elementary. It may even appear you think they are dumb. You should assure them that the things you are doing are being done solely to enhance communication and field performance.

With those thoughts in mind, let's consider some important principles you need to keep in mind at all times in developing good communication. First, help yourself and your employees to practice "active listening."

More problems and mistakes have happened in this business, which never should have happened, simply because people misunderstand what was said. Actually, it is not important what was said but what the other person perceives you said.

Work—and help your people work—to listen well. One of the major problems in business is that everybody is talking, but too few are listening, really listening. You and your people should learn to attend completely to the one who is speaking. True listening requires us to hear another's words and register them in our minds. Too often we appear to be listening to a person, but we are only waiting for them to stop talking so that we can present our argument. Or, we are looking for "ammunition" to use against them. This kind of practice will soon cause the other person to become very careful and evasive in what they say, and destroy your ability to communicate with them.

One of the basic practices of active listening is to repeat back to the person exactly what you have perceived they said. It may seem too elemental, but this simple practice can save you thousands upon thousands of dollars a year. Talk about this practice with your employees when you are setting up your communication plan so they will understand why you want them to do this.

Second, make it a part of your communication plan to write down all important communications—after having decided what will constitute important communications. So many times important things are forgotten or misunderstood because they are not down in writing. In addition, when they are in writing, your mind is free to do more important things, such as creative thinking and planning. Also, everyone else has a resource they can go back to read and reread if they need to without being embarrassed.

Once you get a reputation for being a listener, you will be amazed at what people will tell you. This knowledge gives you a much better ability to run your business. Some of the greatest ideas, which will make you lots of money, often come from your employees. New techniques and ways to perform better on jobs will begin to surface through the words of those who work for your firm. An important part of any communication plan is to establish a way for ideas, crazy or not, to be heard.

Third, learn to read between the lines. Good communication involves a lot more than just words. It's often the feelings, emotions, body language, and thoughts behind the words that reveal what is actually being spoken. The key to being a good communicator is in learning to read people. This means listening to what they have to say. Then continue to talk with them as you search for a complete understanding of what they are feeling and thinking. Take their nonverbal communication into consideration. Put it all together and you will truly be a good communicator. You must be careful to adhere to the next principle.

Fourth, do not be a reactor. Often, as I watch two people communicate, they remind me of two bare electrical wires being rubbed together creating sparks that are flying everywhere. One person says something, to which the other reacts negatively, to which the other one reacts negatively, and it goes on and on like that until both become very sorry for the things they said. Carefully think through what you want to say. Do not just react.

I have learned a little trick that I want to pass on to you. It has helped me in this area. Before I say something very important to someone else, something that could be misunderstood, I say it to myself, I ask myself, as if I am rehearsing it to myself, "How would I feel if the other person said this to me?" What would I think I meant? If it does not pass by me as understandable, if I would not feel good if it was said to me, then I do not say it to the other person—or, I reword it until it is acceptable.

Fifth, whenever, and as much as possible, be as positive as you can be in all that you say. I am not going to get into a long dissertation about "positive thinking," or of being positive. Others have done a good job of writing books and articles about these principles. However, I do feel that very little has ever been accomplished by negative talk. Most of your employees will respond much better to words of encouragement and to positive words than they will to bad mouthing and negative words. I fully recognize that there are times when it is impossible to be positive, but even in those times, there is a way to present the negative in a constructive and positive manner.

Sixth, use the power of consensus to come up with solutions that will motivate people to produce. Some people seem to think that the best way to settle disagreements is to get into a room, and yell and shout and argue, until one person gives in, thus making the other person the winner. Sad to say, they do not understand that in that type of conversation, there are no winners.

Let me give you an example of why I say that. Let's say that my wife and I are going to buy a new car. She wants a Mercedes 380 and I want to buy a Subaru. We go off somewhere and yell, scream, and present our arguments, trying to intimidate each other. Finally my wife gives in and we go buy the Subaru. Have I won? Not on your life! Over time as she gets into that car she will hate it. If it ever has mechanical problems, she will remind me that we bought the wrong car. Here is how we should have decided what kind of car to buy. She comes off her Mercedes to a Saab, and I move from my Subaru to a Ford Taurus. We both move from those cars to an Eagle, and then go out and buy one. That way we have bought something that was neither one of our exclusive ideas, but rather something that we can both accept. That's making a decision by consensus. I have often found that decisions made in this way are better decisions because you have used the good ideas of someone else to temper your own ideas. You are now getting the best of both ideas.

Too many owners feel that this kind of approach will call their authority into question. If you are going to take this approach, you must reserve for yourself the first authority to override any decision. However, nothing should stand in the way of allowing your people from being a part of fine-tuning your ideas through consensus.

Lastly, nothing should be too small to discuss. I find that there are many big problems that come up in a company that did not start out that way. They started out as small, apparently insignificant problems, but were not addressed when they were small. Because the owner (or management) did not know them and did not deal with them while they were small, they began to get bigger. Soon they became massive problems that cost thousands of dollars to correct. And the sad thing is, they could have been corrected for just a few dollars if they would have been handled when they were small. Nothing is too minor to discuss in a good communication plan. It is better to have to listen to a lot of little things that do not matter, than to overlook that one little thing that later becomes a big problem.

Finally, the success of any communication plan is to make sure that you — and it — are flexible. Nothing should be carved in granite. Everyone should remain open to any new or better way of communicating. Also, everyone should realize that you will never attain perfection. Everyone must continue to strive towards improving the communication in your company. Too many people give up too quickly because they expect perfect results at once. Your goal should not be perfection, but rather the continual improvement of the communication in your company.

Starting and Maintaining Job Costing

A contractor without a job costing system is, in essence, operating in a vacuum. He or she will estimate that things are going to take a certain length of time to perform, but then they have no way to determine if it took that long or not. They may then go on to multiply their errors by continually using performance times that are inaccurate.

Many contractors do not have a job costing system because they feel that it is too difficult to implement. Some of them may have attempted to start one at some time but, because they did not go at it in the right way, they failed. Now, because of the hurdles that must be overcome, and because of the bad experience that they had, they give up on having a job costing system.

A contractor should have two systems of cost control. One is done every month along with a monthly statement. That is called *cost accounting* and that will be covered in another section. The other is called *job costing*. It is a challenge to start and manage job costing in labor, but the information you will have in hand will be worth the effort down the road. With that information you can go on to figure the costs of material, equipment and subcontractors.

Getting information from the field can be difficult. Whenever contractors ask field people for certain labor functions for costing purposes they tend to squirm. They think that you are either checking on them or that you don't feel they are doing a good job. Many times, the figures you asked them to keep track of either never show up or, when they do, they've run the paper across a hydraulic cylinder on a tractor somewhere so you can hardly read it.

Consequently, the best method to get those cost figures on a regular basis is to tie it into the payroll. Everybody wants to be paid. So, if a contractor can develop a time card that must be turned in properly in order for them and their crew to get paid, the contractor will get the necessary figures.

On page 157 you will find a sample of a daily time card geared toward computer job costing. There are a couple of features that need to be pointed out and explained concerning the time card. On the far left side are spaces for the project superintendent—or foreman over a particular function—to list all the people and equipment on the project for a given day. That is his role to keep time for those people. If a contractor puts individual time cards in the hands of all workers, the chance of getting back accurate numbers and breakdowns are slim to none. Somewhere on the time card there should be a place for each person to sign by his time, thus showing his agreement with the hours.

Next to each person's name are several blocks in which each worker's hours are broken down by work functions. Before the job is started, the estimator should break down the project labor into work functions that they want to cost and track. Each function is then given a number before the project begins and a master list is given to all responsible persons. For example, a function list might look like this:

FUNCTION NO.	DESCRIPTION
101	Grading
102	Irrigation
103	Large Trees
104	Small Trees
105	Shrubs
106	Ground Cover

The person filling out the time card inserts the function numbers that they worked on that day across the top. Then they fill in the hours each person—and each piece of equipment—worked in each function. On page 157 you see the time card filled in. Notice on the right-hand side a daily total can be kept which will accumulate on a computer so for example, at the end of the week a contractor can total up Jim Smith's or Chuck Casper's total time for the week. They can also add the totals across the bottom to determine the hours spent on each function. Then, at the end of the week, they will have a total of the hours spent that week on each function.

The function numbers and the bid hours were established by the estimator on the original bid sheet. The hours spent come from the time cards, which have been turned in so far for this week. The hours to date come from a running tally of the hours spent to date. The percent complete column is a number supplied by the project superintendent each Friday. At the end of the work day he or she goes over the job and gives their best and fairest estimate of how complete each work function is. Is it 10, 33, 55 or 85 percent done? (For larger items it is best to take the previous week's report and add a percent to it; e.g., did someone do an additional 5 or 10 percent from last week?) These percentages are turned in on Friday, punched into a computer along with all the payroll information, and appear on a report.

The report will help contractors control jobs by telling them if something is going wrong in a work function. It will tell them within a few days of the problem, rather than 30 days later when their books have come back and it's too late to do anything about it. On Monday morning they can be out on a job talking about a problem and trying to resolve it. There are contractors who are doing a lot of small jobs that only take a day or two and who get this report from their computers on a daily basis.

At the same time, this form sets an attitude in the field people. They know how quickly they need to get something done before it will start costing the company money. These numbers can become a measurement of their ability and they can find great pride in getting something done more quickly than was estimated.

But this report also gives contractors up-to-the-minute estimating costs. If something last week and the month before took their people longer to do than they estimated, they should not use those same costs again this week on a bid. They should raise their labor production figures to correspond with what is actually happening in the field.

The whole process of using past labor production times on future jobs takes a personal touch. It takes a person who understands construction and can understand how a cost equates to the work done, and how and when it can be used again. An estimator needs to be able to say, "This job I am estimating is almost like that job we did, so I can use those costs and alter them a little for this job...." That's adding the personal touch.

There is a company that almost went broke over this very thing. When computers came out they were picking up work like crazy but going broke just as quickly. Their computer people were simply sticking costs from one job into another estimate without the personal touch. They were estimating jobs in rock, for example, using costs from jobs that were done in sandy soil, and so on.

Let's talk now about some of the common problems faced in starting job costing and some of the possible solutions to those problems.

Often field people do not properly perceive the purpose for a job costing system. Others may feel there is some ulterior motivation on the part of management in starting one that will end up costing them their jobs or their future pay raises. Consequently, because of these misperceptions, they may sabotage a contractor's attempts at starting job costing. Thus, the first step to starting and then maintaining a good job costing system is to effectively communicate why one is needed in the first place, and then to be able to sell the idea to those who will be doing the record keeping in the field.

There are three things that a contractor must communicate—and sell—about job costing if it is to succeed in their company. First, job costing is necessary for the estimator to know what he or she is doing. Field people often think that enlightening the estimator is the last consideration of management and that every time there is a difference between estimated performance and actual performance they—the field—will be blamed. It would be preferable to let the field people know that job costing needs to occur primarily to educate the estimator.

Second, job costing should be viewed as a motivator. Every one of us needs motivation. Any person who says that he or she is completely self-starting, and does not need anyone or anything to motivate them, does not truly understand himself or others. Everyone needs motivation of some kind to be the best he or she can be in life. A job costing system needs to be viewed by everyone as the best motivator for people in the construction business. It should be viewed as a goal to be attained, a target to be hit, a schedule to be met, and a budget to be achieved or even beaten.

It is also the system by which people are informed about what must be done and how quickly it needs to be done. An old proverb says that without a vision people perish. Job costing is the vision that enables construction people to know what is expected of them. If you are given a job to do, and you are told to do it as quickly as possible, there is no motivation for you, because you have not been told how quick is quick. On the other hand, when you are told exactly when you must complete the job, you have been motivated by giving you a precise goal that enables you to know when you have successfully accomplished a particular job.

The team effort of both the estimator and the field people in estimating is very important. Every estimate that goes out of a contractor's office ought to have the input of the field people. Job costing, then, can refresh the memory of the field people as to how long it took to do a certain job function on previous jobs that were similar to the one currently being estimated. If a contractor can communicate and sell that as a primary purpose for job costing, field people should be more willing to keep the records in such a way in order to be able to understand them at a later date.

Job costing should also be seen as a way to measure performance. Now, this is the very reason some field people do not like job costing. They feel that they do not need to be measured, that they do a good job, and that we should just accept that at face value. Again, let's turn from the negative thoughts that enter people's minds concerning job costing, and instead turn to the positive view of why it needs to be implemented.

To those who have that kind of outlook, let's propose the following. Yes, we think—as you think—that you are doing a good job. But, job costing gives you the opportunity to prove to yourselves and to management that you are doing a good job. With job costing, a field person's good performance is not left up to people's subjective views of whether or not it is good, but is now put into an arena of understandable and unbiased facts. This will benefit greatly those in a contractor's field who are good—and they will all think that they are—and will weed out from the company people who are giving the good people a bad name and jeopardizing their job security.

Again, the success of starting and maintaining a good job costing system begins long before it is actually instituted. It begins with management communicating and selling to the field the great benefits that this system will bring to them and the company.

Another reason job costing systems sometimes fail is because some contractors try to do too much too fast. Remember, one has to learn to crawl before he walks, and to walk before he runs. It is the same with job costing. A contractor cannot expect to have a fully operational job costing system in a month. In fact, it can take up to six months or more to have it working in such a way where he will begin to reap the benefits. Many contractors are not willing to wait that long and so they "trash" the system before it really has a chance to work. Some feel that since it is not working within four or five months, it will never work with their people.

They do not realize they may be very close to having things fall into place. Other contractors, for instance, may end up putting so much pressure on their people about it and it becomes a point of tension. Then they feel they will eventually lose good people if they continue to pursue the establishing of a job costing system.

All of these problems can be avoided if, from the outset, everyone involved in management knows it will take some time for the system to become operational. An old saying goes, "Good things come to those who wait." The good thing of job costing will come to a company, if management gently persists at it, and waits patiently for the day when it works.

Another reason some companies fail at their attempt to start job costing is because their system becomes overly complicated. Many times, a company going into job costing has not even kept track of who was on what jobs, and for how long. They merely have a weekly time card for paying all their people. They should start by simply asking people to keep track of who is on what job, and for how long. Once the field people gain a level of confidence with that, and discover that is was not too difficult to accomplish, a contractor can move them on to the next step, asking them to keep track of the time spent on the job in two or three different functions of work.

Let's say that the job has electrical, plumbing and carpentry that they are doing with a contractor's own work force. Simply ask them to keep track of who worked in those major areas and for how long. Once they feel they can do that, then begin to ask for even more details. It may take six months to a year to get from the first step to where they are supplying the kinds of detailed job costing a contractor wants. But, this is the best—and the least frustrating—way to get there. Some companies come out from day one with a hundred or more things these field people—who are brand new to job costing—must keep track of. As a result they supply a worthless mess of bogus figures shich come back from a frustrated and confused field.

Therefore, start slowly and build people's confidence, letting them get used to the ideas before you demand they be able to do job costing in all of its fullness. Getting job costing started, however, is only half the battle. Keeping it going is the other half. The field people must see benefits of the record keeping from their standpoint. If job costing is done and they do not know someone is looking it over and using it, they will feel it is a waste of their time. If a contractor does not make comments—good and bad—about the field's performance based on job costing, then they will begin to wonder if keeping the necessary records is worth their effort.

Also, a contractor must resign himself to the fact that in order to get good information he must become like a broken record. He must continually "get after" people. He must continually sell his people on benefits of job costing. No system which involves people will work well long term without this kind of constant encouragement and motivation.

Job costing is the first step in monitoring and getting field performance, and acquiring good production hours to use in estimating. Get your system started now!

Project # 8651
Project Name CENTRE COURT

Date: 8/7/85
Foreman: Bill R.

FULL NAME	TIME IN	TIME OUT	TIME IN	TIME OUT	TOTAL HOURS	101	102	103	104	EMPLOYEE INITIAL	TOTAL
1. Bill R.	8³⁰	12³⁰	1³⁰	5³⁰	8	2		6			8
2. Jim S.	8³⁰	12³⁰	1³⁰	5³⁰	8	4	4				8
3. Chuck C.	8³⁰	12³⁰	1³⁰	5³⁰	8	4	4				8
4. Susan R.	8³⁰	12³⁰	1³⁰	5³⁰	8	2	2	4			8
5. Ralph M.	1-		1-	5³⁰	4				4		4
6.											
7.											
8.											
9.											
10.											
TOTAL HRS ON PROJECT					36	8	10	10	8	TOTAL	36

EQUIPMENT

DESCRIPTION (Your Equip)	RENTAL	TOTAL HOURS	103	104		TOTAL
1. Bob CAT	6		6			6
2. Fork Lift	4		4			4
3. Pick-up	8		4			8
4. Two Ton	8		4			8
		TOTAL HOURS ABOVE	103	104	TOTAL	

SUB-CONTRACTOR / OWNER-OPERATOR

LIST NAME	TIME IN	TIME OUT	TIME IN	TIME OUT	TOTAL HOURS	Describe what was done/down time/etc.	TOTAL HOURS ABOVE
Bill's Backhoe	8³⁰	10³⁰			2½	Dug Holes On South Side	

NOTE: ☐ Materials Delivered ☐ Materials Needed ☐ Progress ☐ Problems ☐ Visitors, Etc. on Back

Example of daily time card geared toward computer job costing.

Cost Accounting Report

Job Name: CENTRE COURT   **Job No.** 8651

For the Week of: 8/5/85 **to** 8/12/85

Function	Bid Hrs.	This Week	To Date	% Completed	Gain (or Loss) to Date	Extended Gain (or Loss)
100	100	60	120	90%	-20	-32
101	150	80	120	80%	-0-	-0-
102	50	40	40	90%	+5	+9
103	80	40	40	30%	-16	-64
104	200	60	100	50%	-0-	-0-
105	30	-0-	-0-	-0-	-0-	-0-

Cost Accounting Report Example

Control and the Appearance of Control

Blue collar theft is, by far, a greater dollar drain on the American economy than any other form of theft. Construction companies, with their many jobs, great quantities of materials, and difficult security problems, become one of the largest victims of this kind of crime. Many companies are nothing but a memory, simply because they were so terribly ripped off by their own field people.

This is largely due to the fact that in construction we are sometimes forced to hire less-than-scrupulous people. But this is certainly not the only reason for theft. In fact, many times, theft is committed by some of our very good people. It often happens because they feel as though it is an accepted, or even expected, practice by those in this industry. Sometimes they justify it because they feel that you owe it to them and, even though you have not given it to them, you would not mind them taking or borrowing materials and equipment.

I think in order to keep this kind of thing to a minimum, you need to always work at letting your field know that you are in control. If you let them know you are on top of things, it will help keep them from being tempted to take, or to use things that are not theirs. In working out how to do this, I think you need to understand the difference between control and the appearance of control. You need to decide first how much time and effort will be given to control, and then second, to decide what can be done without spending a lot of money but still giving the appearance of control.

Control will happen in those areas in which you spend time, money, and effort to keep track of important items. Keep on top of those things and you'll have control. Everyone in your company should know that you are aware of everything about those particular items and that no funny business can occur in those areas without you knowing about it immediately. That is the level of control we all need to keep from being ripped off.

But we all know that in this hectic, fast-moving business, it is virtually impossible to have that kind of control in every area of every job that is going on in your company. In some cases, it would actually cost more to keep people from stealing things than the things they would steal are worth. It is in those areas that I feel you need to come up with some inexpensive ways to give the appearance of control. Let's list some areas in which you can have problems, and then discuss some ways to put forth an appearance of control to keep these areas from being a problem.

1. Ghosting: The practice of putting friends or relatives on the payroll when they are really not even working for your company. They then collect a paycheck and split it with your employee.

2. Padding hourly contract people (truckers, backhoes, tractors, labor): The proactive practice of paying someone a few dollars more than they want or giving them more hours on their job than they actually worked, then splitting the difference with that contracted person.

3. Tool and material pilferage: The practice of taking items from the job and either using them on their personal property or selling them to others.

4. Equipment borrowing: The practice of taking your equipment home on the weekend and either doing work at their home, or doing work on jobs they have contracted for personally.

5. Drug use before the job or while on the job.

6. Floating: The practice of taking too much time for personal business away from the job site, or of letting the crew people take too long on their breaks, or of not producing a fair amount of work in one day.

Now that we have listed some of the potential problems, let's talk about ways to prevent them from happening. First, ghosting employees. I think that it is important that the owner not only sign the paychecks, but that on unannounced occasions, he goes out to the job site and actually pass out the paychecks and see these people face to face.

Second, padding hourly contract people and floating. I feel that all hourly prices should be negotiated by people in management positions and/or the owner, rather than allowing field people to negotiate those prices. Also, field visits should be made at unannounced times. These field visits will ensure that people are actually on the job. Also, these visits should include one well-placed question about how long someone (contract hourly person) has been on the job and is expected to be on the job. That information should be written down in the person's presence even if you never look at it again. Also, I think that the best source of information can be other people on the job other than your foreman or supervisor. Again, you do not want to appear that you don't trust your foreman or supervisor. You are merely seeking information that will give you what you must have; i.e., control over your company. Even if the information is not used, just the act of collecting it will help keep things under control.

Also, job costing both labor and equipment is one of the best ways to control these items. If these items overrun the original estimate, some sort of justification must be made by the field supervisor and/or the estimator. However, I must admit that job costing can cause you a problem in this area. If someone sees that more hours were estimated than were needed, they may use the extra hours needlessly, which is theft, really. In the next section,

I will talk about rewarding those who do well and who do better than the estimates on a regular basis. If you do not do this type of thing, then you will set yourself up for the kind of problem I have just mentioned. Consequently, rewarding people for a job well done is another way to minimize theft in these areas and is a must if you do job costing.

Third, pilferage. This is probably the most difficult thing to control on your jobs. One of the best ways is to have job costing done in the area of material. However, this can be expensive to do if you are a smaller contractor. I believe every contractor can do job costing for labor and equipment, but only those who are bigger, and who can afford it should do material job costing.

But regardless of whether you do material job costing or not, every company should have someone (preferably the owner) look over the invoices and, again, ask certain questions of certain people which will always give the appearance you are in control of what material goes to what jobs. Also, this same person should make spot inventory checks (on the jobs) for the materials which were ordered and for what was actually installed.

One other system I recommend is a two-type purchase order system. I recommend a large-type purchase order be used for the original purchases of material the estimator had taken off and felt was needed (right from his estimate sheet). Then, all field people should be given another, smaller purchase order with a different numbering system. These are to be used by field people whenever they need to order or pickup something in a hurry. Anytime one of the purchase orders is used, a red flag should go up in your mind. It means that the estimator missed the quantity on his estimate, or that something has been stolen, or possibly, a change order needs to be issued.

Though it can be a hassle, all your suppliers and employees need to know that nothing can be purchased over $25 in value by anyone in your company without a purchase order. Your attention to such details will help you eliminate pilferage.

Fourth, borrowing equipment. Many times, this type of thing can be acceptable practice within your company. But if you do not have some policy as to how and when an employee can borrow a piece of equipment, it can get away from you very quickly. And, even with a policy, some equipment can be borrowed without permission. Again, job costing as to the hours certain pieces of equipment are used on jobs can help reduce this problem. When you go on jobs, it may be wise not only to look over equipment to make sure it is not being abused, and it is being properly maintained, but also to write down the hours on the hour meter, or the mileage on the speedometer. Remember, this information may or may not be used by you, but the appearance of your checking and monitoring these things can help reduce any problem in this area.

Finally, drug and alcohol use before or while on the job. In talking about this problem with companies around the country, I find more and more firms are taking a very tough stance concerning this problem. Some are not allowing any kind of drinking on the job, even after the day is over. They are worried about potential law suits if one of their employees is in an accident after what could appear to be a company-sponsored (or condoned) time of drinking. Also, they feel they want to make it quite clear, from an easily understood policy, there is to be no drinking on any job at any time.

Also, many companies are instituting mandatory drug testing prior to any person being hired. This does not completely insure you of having a drug-free work force, but it does send a message that is heard loud and clear by everyone. Be sure to consult with your attorney about establishing and carrying out this policy.

Remember, good field performance, and honesty in what is going on out there on all your jobs, depends on how in control you are. Make every effort to control those things which you can, feasibly and economically. For those things which you cannot control as you would like to, find ways to at least give the appearance of control to your people.

Job and People Scheduling

Nothing has an impact on the performance and profitability of a contractors field more than the initial scheduling of material delivery, job start time, and the right number of people needed, to be most productive on any given day. Contractors need to spend a lot of time planning a job correctly in order to make it go right and end right.

The first problem is jobs which are started earlier than they should be. This is usually done for one of two reasons. Either the contractor is low on work, and so they shift people onto a job before it is ready, or they are forced to start a job early because of an owner who is making unfair demands.

However, in certain situations, starting a job early is justifiable. It is justifiable under the following circumstances:

1. A contractor has a lot of work, all of which is to start at the same time in thirty days. The more they can get done on any of them now, the fewer the problems they will have in the bottleneck which will come in thirty days.

2. Even though they are starting early, the things they will be doing can be done within 80 percent of the original production times.

It cannot be justified in situations where a contractor does not have additional work (or the hope of additional work) that provides pressure, causing them to get as much done on any job as they can. If a layoff or slowdown is coming, why not do it before a job? A contractor can then maximize the profit on the job when the time comes to do it right and the most efficient way.

Everywhere contractors have the same problem: owners wanting them to get into a job before it is ready. The nature of this business requires them to take charge of their company and its jobs. It will require them to take this approach: the owner is changing the normal condition of the job, and they must either wait for the right startup time or accept a change order for additional costs. So often, contractors do not ask for these kinds of things in a matter-of-fact way, and instead just do as the owner requests, while complaining about the problem. A contractor will not get a change order every time, but the ones they do get will certainly help.

The second problem is where jobs are often not planned properly. This happened in several areas. Sometimes, material shows up at the job too early or too late. Other times, not enough men, or too many men, are put on the job. Still, other times, the proper equipment is not on the job when it could have been done better by hand, or vice versa. All of these things can be avoided if more planning goes into the job.

Planning needs to be done in two ways. First, if a contractor has a good estimating strategy and system, some of the planning has already been done during the bid process. Also, the estimate will tell them how many hours it will take to produce each proportion of the work. Total the amount of hours which has been estimated for each proportion. Decide how soon it must be done to fit the contractors schedule and the schedule of the owner. Take this number (of days) and multiply it by the number of hours estimated, and it will tell you how many men to put on this portion of the work to meet the schedule. Then decide if the number can work efficiently. If not, you may want to change the schedule.

Here is an example. Let's say that a contractor has estimated 220 man hours to do a portion of work which needs to be completed in 4 days. The field people will work 9 hour days. Consequently, multiply the 4 days by 9 hours and the result is one person would complete 36 hours of the work in 4 days. Then divide 220 man hours by the 36 and it tells you 6 people would be needed to get the work completed in this time. Then decide if 6 people can work efficiently on this job, or whether it would be wiser to cut it back to 5, or even 4. You can quickly see that with 4 people, at 9 hours a day, 36 hours would be performed per day. If it will take 220 hours to do the job, divide the 220 hours by 36 hours (per day) and discover with 4 people it will take 6 days to do the job.

Also the decision on whether or not to use certain equipment is a very difficult one. It often becomes simply a judgment call. However, there are some things to be considered. First, will the field people be less "worn down" if equipment is used? Is there a shortage of good people, so anything done with equipment is helping your people shortage problem? Second, do you have the equipment anyway and if it is not used, it will just sit and collect dust? Third, and most important, is the cost of the equipment and what it will produce per hour, more cost effective than people and their cost per hour (including labor burden)? Fourth, is there pressure to finish other jobs and this one needs to get done as fast as possible, even if it is not as cost effective? Again, there are no hard fast rules; it is very much a judgment call on almost every situation.

The scheduling of equipment and material requires a contractor to be constantly thinking and looking ahead. Contractors use many different kinds of systems to accomplish this need. Some use calendars which have huge spaces for each day. They use the space to make notes about everything which must be ordered, or moved in, on any given day for each job. Some use planning boards with notes pinned in appropriate places. In the next paragraph is an example of such a board. However, it does not matter what is used, as long as one uses something which puts it in writing. Too many contractors are trying to do all of this work in their minds. The mind should be used to think of new and better ways to get jobs done, not to store information. Once one has a thought, or sees a need coming in the future, they should put it on paper, then put in a place where they will come across it in time to take care of the need.

On the next page is an example of a job scheduling and planning board. It simply is a two week calendar with spaces large enough to list all the foremen and major pieces of equipment. This board will be used to initially decide where key people will be, with a crew, on what days and with what equipment. Once this is filled in, a form like the one that follows the job scheduling board should be filled out for each individual. It has more detailed information about that person and the job. From this form a contractor can make notes as to when to order materials, and when to have equipment moved from one job to another. Again, the key is not the shape of the form, but that a contractor has a form to use in order to schedule everyone in a general way, and then a form to take care of the particulars for each individual job.

How would you like to make $100 to $1000 an hour? You can! Every hour spent in planning, and in maximizing one's people on each job, can easily produce that kind of income for a company!

INDIVIDUAL JOB SCHEDULE AND PLANNING BOARD

	MONDAY	TUESDAY	WEDNESDAY	THURSDAY	FRIDAY	SATURDAY
JOB						
PEOPLE						
EQUIPMENT						
MATERIALS						
MISC.						

Individual Job Schedule and Planning Board

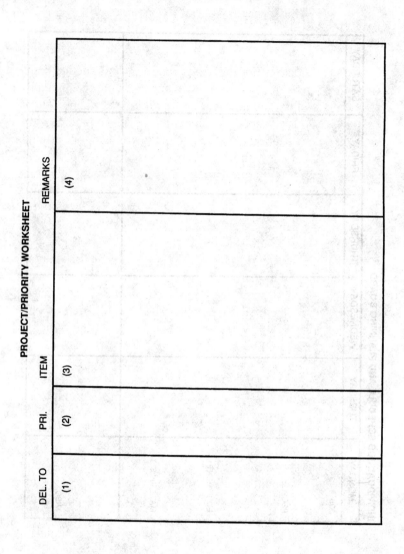

PROJECT/PRIORITY WORKSHEET

DEL. TO	PRI.	ITEM	REMARKS
(1)	(2)	(3)	(4)

Project/Priority Worksheet

Management by Teamwork

One of the things I believe in very strongly is construction companies need to be run by teamwork. If you are running your company dictatorially, you are running it with tunnel vision. There are things that are happening to you which you cannot see, and you will never see, because you are not allowing anyone around you to show you those things. I want to describe to you a team I was on which ran a very large site-development general-contracting company for several years.

There was something almost magical about this team. We had a camaraderie that would not quit. We also had mutual respect for each other and for our roles within the company. We understood one of the most important human dynamics: We understood all of us were strong in some areas and weak in other areas. Consequently, we covered each other's weakness with each other's strengths. Where I was weak, someone else was strong. Where I was strong, someone else was weak. And, the success of our team was due to our willingness to recognize this fact and to work it out in our working together. There was one other important ingredient this team had -- no one of us was over another one of us. We all felt an equality in our special roles of making contracting firm successful.

Let's consider the roles each of the three of us had to fulfill. First, let's consider the field superintendent. He was the one who saw to it the work got finished. He coordinated all of the people, the materials and the equipment onto the jobs. He worked out the methodology which would be used in getting the job completed. If there were problems on the job, he got the necessary people involved to get them resolved.

There are a couple of personality traits I think are important to a field superintendent. One, they have to like—and be able to motivate—people. What is the biggest risk in construction? Labor. Where can you make the most money in a consistent way in construction? Labor. And, a field superintendent who likes people and is liked by others, and who knows how to motivate people, is going to make you money. Two, he has to be the kind of person who doesn't mind problems and, in fact, is challenged by problems. There is no greater place for problems than in the field, so whoever is in charge had better not get frustrated with handling them. The person with whom I was on a team was just such a person. In fact, to this day, I accuse him of starting problems, if there weren't enough around, just so that he could keep himself challenged. If he was a fireman, he would have been an arsonist.

The second person we need to consider is the accountant. On our team, this was a lady and she was a good one. She didn't take any jive talk from any of us. She could read right through some of our talk. She kept all of the books, did payroll, the accounts payable and receivable. She was also the office manager in charge of all the secretaries and the flow of paperwork around the office. I was the third person: the contract administrator. I first decided what to bid using the criteria I have written about in the section *Troubleshooting the Estimating and Bidding Process*, with one addition. I would always bring in the field superintendent to see what he thought about a job before I bid it. If the job made him sick and he didn't like it, or was not challenged enough by it, we didn't bid it. If we had bid a job he didn't like, and got the job, he would still be sickened, and you can just bet that it would not have been a very profitable job for us to do. If we were low bidder on the job, I would negotiate the contract, if needed, and get it signed. I would see to it that all the preliminary paperwork was done; that is, the pay estimate forms were approved, submittals, work flow charts or CPM charts were taken care of; etc. I would then set up a preconstruction conference with the owner. However, before we went to the preconstruction conference, we had one in our office. We wanted to get all of our ducks in a row before we met with the owner.

The meeting was attended by the field superintendent, the project superintendent (if one had been designated), the accountant, the owner, myself and Mr. Tape Recorder. The meeting was taped and later typed out. It then became a part of the job file and the project superintendent had a copy to refer to continually when he needed to know what we had talked about and agreed upon during that meeting.

One of the first things we did was to review my estimating mistakes and see how we could minimize their effect on the company. We looked over the job with one goal in mind: finding a way, some way, to make that job better for us. I remember being low bidder on a highway job in a large city in our county. On this job there were several median islands that had very bad soil in them, and they wanted the soil replaced with topsoil mix for plants. We were to excavate the bad soil 18 inches and replace it with topsoil. There was only a pay item for the new soil. The removal of the old soil was to be bid in with putting in the new. There were 10,000 cubic yards of material to be taken out and replaced. On the same job we were to build some huge earth berms. They required we bring in 30,000 cubic yards of material. I had gotten a quote of $4.50 per yard for the imported dirt. At the same time, the trucker quoted me a figure of $4.50 to haul off those 10,000 cubic yards of bad soil. This is the way I figured the job, $4.50 per yard for new soil and $4.50 to haul off the old. We were sitting in a pre-construction conference among ourselves in our office trying to figure out a way to make this job better. Someone asked, "Charlie, would they let us take that old soil and put it in the bottom of those berms and just import good soil on top?"

"Let's ask them," is all I could respond.

And so, at the end of our pre-construction conference with the owner, we asked them. They blinked their eyes a few times and said we could do it that way. We put $90,000 back into our pockets at that very moment! We did not have to haul off 10,000 yards at $4.50 per yard, and $4.50 per yard they paid us to import what we did not have to buy. That's 10,000 yards at $9 per yard, or $90,000! And, it was all because we were willing to ask ourselves, "How can we make this job better?"

I wonder how many of you reading this book are wasting thousands of dollars on your jobs right now. Oh, I didn't say you were not getting your jobs done for the cost at which you bid them, but you could get them done for a cheaper cost. You could come up with some new methods or approaches, some other pieces of equipment to use, or a different material which would give the owner the same quality of job but save you money. And, the only reason you are not doing it is because you didn't sit down before you started the job with some of your key people and ask the question, "How can we make this job better?" You see, once you sign a contract, the price is set. The only way you can increase your profit is to lower the job's cost to your company by finding a way to do it faster and cheaper. We took it on as a personal challenge to find ways to make jobs better. I always say you need this before you ever start, because once you start, the job will find a way to make itself worse.

During the meeting we also decided on what subcontracts we would award and where we would buy—or shop—for material. We laid the job out as to its expected progress and billing schedule. We also pinpointed potential problem areas and began to lay preliminary plans as to their solution in our favor.

After the meeting, I turned the job over to the field superintendent to be done. However, I wanted to be involved in it in three different ways. First, I priced all change orders. Don't let your field people price out change orders. They will do it on the back of some matchbook cover and burn your whole company down. I knew what was on the job and how it had been going and how the architect/engineer had treated us to that point, so I wanted to price the change order. Also, as I have already mentioned, change orders are dangerous, so I wanted to be part of estimating their true cost. I also was involved in the monthly billing process. I would work with the field superintendent to get the pay estimates figured and submitted. I wanted to do it so I could stay in touch with the reality of how quickly that job was really going together. I also looked over the payables to get a feel for the true costs which were being spent to do the job.

Let me insert here a suggestion as to how I feel the payables of a construction company should be handled. I feel the accountant should approve them for math. Does the quantity multiplied by the prices work out mathematically? Then they should go to the field superintendent to approve for quantity and quality. Did you receive the amount of material and was it acceptable? Then they should go to the estimator/contract administrator for them to approve the price. That person bid the job and should be aware of the price quoted them at the time of the bid. Often, I have found suppliers adding delivery charges even though they had quoted me a price FOB job site. Sometimes, they charged a higher price because someone in their firm didn't know of the special price which had been quoted for the job. After an invoice clears this process, it is ready to be paid.

There is one final thing I liked to be involved with and it is the final "punch list." Don't ever let your project superintendent do the punch list. I have found they can give away the farm just to get off a job. Let's say that person has been out there on the job for months and they are sick of the job. The owner and the architect/engineer have given them problems day after day. Subcontractors have given them the run around on a daily basis. At this point, your project superintendent will do anything to get off the job—then here comes the architect/engineer to do the punch list.

They quickly read your project superintendent's attitude and know they are in the driver's seat. They can ask him to do things which were never even called for in the original documents, and he will do it just to get off the job. He may be willing to write the architect/engineer a check just to get off the job. I, on the other hand, could come out there a little less frustrated and a whole lot fresher and be able to hold my own better with the architect/engineer. Thus, I have found I could do a better punch list.

Let me add one final bit of advice. Remember this principle: the more quickly you can get in and out of a job, the more money you are going to make. Not just from the possible labor savings but from many other avenues as well. A large construction company recently tore down a major viaduct into a downtown area of a large city under a 13-million-dollar contract. They were given 400 working days to do the job with a liquidated damage penalty of $10,000 for every day they ran over on the time limit. They were also offered a $10,000 per day bonus for every day they finished earlier than the 400-day time limit. They finished the job in 300 working days and there were pictures of them collecting a $1,000,000 bonus check in every major newspaper in that city. Is that all they made by finishing that job 100 days early? Not on your life! For 100 days they were not paying out general condition costs for trailers, temporary toilets, job phones; etc. For 100 days they had one dynamic project superintendent—who could build that viaduct in 100 days less than anticipated—over working another job, which would make them more money. And, when it came time for the punch list, the owner was so thrilled to have the job done early, I'll bet the punch list wasn't very long at all.

However, if you stay longer on a job than you should, the opposite begins to happen to you in a big way. Your costs for general condition items begin to exceed your estimate. Everyone's attitudes on the job go downhill, and the production begins to go bad. You have a project superintendent who is still on that job losing you money, plus losing the money he or she could have made you on another job during that time. You have an owner and an architect/engineer who are mad—and they are mad at you. When it comes time for the punch list, it's not six items, it's six pages of items. I have heard contractors take on the attitude that it takes a certain amount of time to do a job, and they will finish when they finish. This kind of attitude costs thousands of dollars. Remember, the more quickly you can get in and out of a job, the more money you will make. For that reason, I like contracts with designated completion times and liquidated damage clauses. They force me to make money!

Management by teamwork, a much better way to run one of the most complicated businesses around: the construction business.

True Life Stories From The Real World

✦ *I Found The Grass Isn't Always Greener*

Before I entered into my own business I was fortunate to have a good informational background in the industry. I was the office manager for Vander Kooi and Associates, Inc., and had assisted him as he conducted seminars and workshops. I had ample opportunity to be exposed to and learn his principles.

By adding my own personal experience to this book, I am not going to try and sell you a product, I am not going to glamorize the life of a contractor, nor am I going to try and tell you how terrible the life of a contractor can be. I am just going to provide you with my side of the experiences which I have had in the past in the landscape business.

It was the spring of 1988 and I had just been honorably discharged from the USAF after nine years of service. The government was getting extremely tight with its money when it came to annual pay raises and I was looking for something better. I decided to move to Colorado and attempt to make a living in the "civilian marketplace."

Filled with high hopes of big money, I began my search for employment at virtually every major corporation throughout the Metro Denver area. Luck was not on my side, however.

Finally in August of 1988, I was offered a job at Vander Kooi and Associates, Inc. as the office manager. While in the Air Force I had learned some extensive computer skills, which Mr. Vander Kooi felt would be helpful in the office.

For the next year and a half, I was given training by Mr. Vander Kooi on the same estimating system and strategies which he teaches to thousands of contractors per year nation-wide. I read all of this books, attended his seminars, and within a year found myself installing estimating programs, conducting the first day of our two-day workshop, and even providing consulting services to some of our clients.

My mind began reeling, I was making it big! Time after time clients would call me and tell me how great things were going because I had helped them, or given them the information they needed to get their company on the right track. Things were going my way.

In April of 1990, my wife and I had a beautiful, healthy baby girl (to add to our two sons), and life was looking cheery. About one week after my daughter was born, I was approached by one of our local clients. He lived just up the road from me and asked if he could stop by to talk to me. I had spoken to him on a few different occasions over the phone, and he had even done some work on the house that I was building. I told him sure, he could stop by.

He arrived quickly, and after a few short minutes of amiable talk, he told me that he wanted to make me an offer. Immediately my curiosity was aroused, so I was listening intently. He told me that he knew the summer months were slow for us in the consulting business, and that he wanted to make me an offer that would keep me busy all year long. I was getting excited now, I think my palms were even sweating! I knew what he was about to say, and he did. He proceeded to ask me if I wanted to become the Vice President of his corporation.

Now I was in dream land, I saw fire works, heard bells, you name it! You see, for the past couple of years, my wife and I had on numerous occasions discussed the possibility of starting our own business, we just didn't have what we needed to do it.

This was our magic carpet ride. This was our ladder to the top. We were going to make the big bucks! Sure, it would take a little sacrifice on our part, but boy would we reap the benefits. You see, Jimmy (not his real name) had only been in business for a few short years. He was excellent at installing irrigation systems or completing a beautiful landscape for you. But he had one major drawback, he knew very little about running a business, and I was going to handle that.

Jimmy told me that if I accepted, he wanted my wife to work in the office, I would handle all of the business end of things, and he would find the jobs and run the crews. I couldn't believe what I was hearing, it was too good to be true. We had struck our gold mine. With my wife working in the office, we wouldn't have to pay for a baby-sitter, he said she could bring the baby to work with her. The local economy was getting ready to boom at any moment, which meant a lot of money to be made, and by gosh we were going to get our share. He then told me that if I accepted, after a period of one year he would give my wife and I 39% of the stock in the company, he would take 51% and ten percent would go into retainage. Now I knew we were going to be rich!

The next day I made an agreement with Mr. Vander Kooi that as of May 1st I would work for him only one day a week, on Wednesdays. This was fine by Jimmy, and I was on my way to becoming the biggest landscape contractor around. Why the biggest? Because I had the secret formula, taught to me by Mr. Vander Kooi himself, and I knew how to use it.

The first few days working with Jimmy were extremely hectic. We had to reorganize the office, set up a working schedule, set up our overhead budget and hire new crews. The previous year Jimmy had done approximately $65,000 in sales, and this year it looked like he would do somewhere in the area of $85,000. But that wasn't enough for me. I knew that we could do much more than that, and I was going to prove it. I asked Jimmy what he wanted his sales for 1990 to be and he answered he wanted to do $200,000. I knew that this was possible, and more, and didn't hesitate to tell him so. He told me that he didn't think there was enough work in the market place to do more than that, but I quickly told him he was wrong. I told him that if he could find the people to do the work, I would make sure he had the work to do.

After two days of scrutinizing the operation, we were ready to go. We hired laborers, an irrigation specialist and a field supervisor. The next day we began a job working at an Air Force Base, and now the money was going to roll in. Or so I thought.

We had two jobs going at one time, the one at the base and another constructing a rest area for the Department of Highways. We had sent the field supervisor and one crew to do the highway job, so it was decided that in order to cut overhead, I would supervise the crew at the base and Jimmy would go do take-offs for me to estimate in the evenings.

After working with the company for just two weeks, I had estimated nine jobs. We were awarded seven of them totaling approximately $200,000 in sales. Our cash flow was not the greatest, so it was decided that we would get a loan using the contracts as collateral. We went to our local bank, which was more that happy to give us a loan. We were going to use the money for payroll until the jobs were completed and contracts paid.

Life was going quite well for several weeks, when suddenly little things began to go wrong. The money that we had decided to use for payroll had all but vanished. And that is when serious problems began. We soon found ourselves in a position of not being able to meet payroll, not a good thing to happen in the middle of a government job.

Approximately two days before the job was completed, I had the "pleasure" of being able to tell our crews that we were not going to be able to make payroll. I told them that we would be able to pay them part (almost half) of the wages, but not the full amount. I was terrified of what the response was going to be, but once again luck was on my side. All of the crew decided that they would go ahead and keep working, as long as we promised them that they would get the rest of their money the following week. We agreed to this for the simple reason that one of the general contractors we had done work for a few weeks back had promised "the check was in the mail," which we knew was more than enough to carry payroll for a few more weeks. Jimmy, my wife and I all decided that we wouldn't take any paychecks for the time being, until some more money came in.

At that same time we began a new job at the airport, which required hiring a few more people. I told Jimmy that I felt that we should quit hiring new people and use the ones we had more efficiently, which would include him doing some of the work.

Work continued as usual for the next few weeks, and suddenly my wife gave me the news I didn't want to hear - the money was almost non-existent again. I felt like I had just run into a brick wall. Not only was this going to cause a problem at work, it was now causing a problem on the home front. We had worked for almost a month and had not received a paycheck, except for the money I had made working one day a week for Vander Kooi and Associates. Jimmy had not taken any money either, but there was a major difference. Jimmy was single, he had no children, and because the company was in his house, it was paying his rent and utility bills.

My wife and I decided that we had to do something about the mess we were getting into rapidly. I called Jimmy and asked him to come over so we could talk. When he arrived, I'm sure he knew that this was not going to be your typical "friendly" meeting. I began by telling him that we were going to have to make some changes in the company and make them fast. I also told him that I felt he would have to start providing some more help in the field. Most of his time now was spent driving around doing take-offs, repairing residential sprinkler systems and other things I was never really sure about. He was very adamant about having a "field supervisor," but I told him that there was no need for one, that was the job I was performing. I had calculated that by having him work in the field, we would be able to save approximately $2000 a month in payroll. He hesitatingly agreed that this was a good idea, and that we would make the change the following Monday. I also told him that unless some money came in REAL quick, I was going to have to seek alternate work. He then told me that he had spoken to one of our general contractors just a few hours ago, and that we were to meet him the next day to pick up a check. Seeing a slight glimmer of hope, I said that would be good and our meeting ended. The next day we did meet the general, did get a check, my wife and I finally received a paycheck and I was beginning to feel a little better. We didn't get full paychecks, but at this point a little was better than nothing at all.

Monday came along, we finished the job we were doing, so I decided that I was going to spend some time in the office catching up on paper work. While there I received a phone call from Jimmy. He told me that he had just spoken to a general contractor in the area who wanted to speak to us about a job which we had bid the week before. It was the largest bid I had ever done, over $350,000, but we had not been awarded the job. A slight surge of excitement ran through me at the prospect of landing a job this large. It would definitely put us back in the black!

Jimmy called me later that evening and said that the general wanted to meet with me the next day, and that he wanted us to do the job. I was crazy with excitement. I had only been with the company for a month and a half, and with this job added to it, had increased our sales from $85,000 to $525,000! I could just see all the money coming in. We could buy new equipment instead of having to rent it, we could get a better office (not in a house) and we could finally hire someone to run the jobs while Jimmy and I made more money. Yes, the grass was really greener on the other side.

The following day I drove over to the general's office, filled with anticipation. On the way over I had a funny feeling in the back of my head. One of those thoughts that fill your mind and bug you all day long, but you don't know why. I called my wife on the radio and had her call the State Highway Dept. to get a breakdown of the bidders on the job and their prices. I then called the general and told him I would have to cancel the meeting and get back with him in a couple of days.

The following day we received the breakdown in the mail from the Highway Dept. and I could not believe my eyes. The general I was about to go see had bid the job at over $200,000 less than the next closest company! After checking over the report, I found our prices had been extremely competitive with all of the other landscape/irrigation contractors, yet his landscape and irrigation accounted for almost ALL of the $200,000. I didn't know what to do at that point. My mind was telling me to forget the whole thing, to walk away from it all. But common sense told me that there had to be a solution.

The following day I drove over to the general contractors office, a big smile on my face. He greeted me amiably and after a few minutes we settled down to the matter at hand. He told me he like my price on the job and wanted to work with me. I said that I was sure we could work something out. He then began to tell me how he wanted me to lower my prices for this item and that item. I told him there was no way I could. He said I would have to lower my prices in order to get the job, my materials were just too expensive. I sat quietly for a moment, pretending to contemplate the situation.

"Mr. Jones,' I told him, 'I'll make you a deal. You supply all of the materials on this job. I will give you a list of suppliers and you set up the accounts. All the billing will be done directly to you and will be your responsibility. Agree to that and I'll see what I can do about my prices."

At first I don't think he understood a single word I said. Then a big smile came over his face and I thought he was going to get up and give me a hug. He shook my hand feverishly and said we had a deal. I was led out his office smiling myself, because I knew exactly what I was going to do. I immediately drove back to my office, took out our estimate and began removing all of the material costs. While doing this, I recalculated our labor hours and labor cost and four of Mr. Vander Kooi's favorite words rang in my ears - The Ding Dong Factor. I was very leery about doing this job for a contractor who had bid it so incredibly low. So I changed the profit from 10% to 12% and increased my labor mark up percentage from 54% to 85%. This job was going to require a lot of coordination between my wife in the office, myself in the field and the State inspectors. The next day I went back to the generals' office and gave him my new prices without the cost of materials and left.

That evening I called Mr. Vander Kooi and discussed the situation with him. He told me my figures would be cutting it close but we should be in pretty good shape. He then gave me a piece of advice that eventually saved the company from total ruin. He told me to tell the general we would do the job under one condition: He was to pay us every week for the first month, then every two weeks after that. I called the general the next day and told him that and he said that was fine.

Meanwhile, work was progressing pretty well. We had just finished one job and were just about to rap up another one. My wife and I hadn't been paid for a couple of weeks, but we were managing.

We were expecting a paycheck that week, so I figured that everything was going to be okay. I was wrong.

The next week my wife told me over dinner that we were in serious trouble. There were bills from the company which had not been paid from over a year ago, and the company was being sued. She also told me one of our former employees was suing us because we had refused to pay for some of his expenses. To make matters worse, we found out our previous personal insurance company was getting ready to declare bankruptcy and had not paid over $3000 worth of hospital bills for us. I felt as if my life had just been shattered. My dream of owning the "big" business had just fallen apart. Reality slapped me in the face and there was no escaping it.

The next day I had a talk with Jimmy. I told him that he had to quit bidding work, and we were going to have to cut every conceivable expense. He began to get angry and stated it was partially my fault we were in this predicament. He stated if my wife and I had paid closer attention to things, we would have more money. Personal pride stepped into the conversation at this point. I began to tell him if he would have listened to me months ago, things would have been different. The day ended with both of us in an angry state.

Later that night, my wife and I had a lengthy discussion about the whole situation. She had already decided that she was going to quit, and she strongly suggested that I do the same. The next day I called Mr. Vander Kooi and told him I needed to speak to him, which I did. During my meeting with him it was decided if I wanted to I could go back to work for Vander Kooi and Associates full time. I was really in a turmoil. A large part of me wanted to keep doing the landscaping because I really loved it and it provided a great amount of satisfaction when the jobs were completed. However, struggling to make ends meet was not the way I wanted to live my life. I talked the situation over with my wife and reached the only solution.

The following Monday the general contractor called us and told us he was ready to sign the contract. Jimmy met him at the job site and was already reviewing the contract when I arrived. I told him I would like to look it over before he signed it, but he said he had already read the whole thing and it was fine. Again that little thought gnawed at the back of my mind, but I said nothing. Jimmy signed the contract and handed it to me. He was happy and thanked me for what I had done for the company. He wanted me to keep the contract with me because I had started the whole project and he wanted me to see it finished. He then walked over to talk to one of the State inspectors at the site. While he was gone I began looking over the contract. Everything was exactly as we had requested, with one item different. Under the payment item, it stated that we were to submit an invoice by 10:00 a.m. on Monday and we would be paid by 12 noon on Wednesday, every other week. It also stated that with the invoice we had to provide proof all rental equipment had been paid for and all payrolls had been made.

Jimmy walked back over to me and began speaking when I began shaking my head. He asked me what was wrong and I knew it was time to swallow my pride and face the truth. I then told Jimmy I couldn't work for him anymore and I was going back to work for Vander Kooi and Associates full time. I don't think I'll ever forget the look of shock on his face. I told him it was nothing personal, but I could not going on working when I didn't even know when I would get my next paycheck. He tried to tell me how great things were going to be because now we guaranteed money every two weeks, but deep inside I knew it wasn't enough. That night I gave him back the company truck I had been using.

I am now back at work with Vander Kooi and Associates, Inc. full time, and life has been a lot easier. Mr. Vander Kooi and I decided that being an office manager wasn't really for me, so we hired someone else to do it. I now spend all of my time helping other companies in their estimating, installing estimating software, and yes, typing this book. All in all I have no regrets about what a disastrous summer it was. I gained a vast knowledge of the landscape and irrigation industry in just a few short months. I can now drive around the local community, and with a great sense of pride, look over several landscapes and say, yes - that place looks great because I installed the irrigation system and constructed the landscape.

Making the decision to quit the landscape business was probably the hardest decision I had ever made. One of my biggest drawbacks (or faults as my wife calls them), is I am an extremely optimistic person. Two of my favorite sayings are "There is a possibility it might..." and "Everything is going to work out fine." Now I have changed to "We need to really sit and think about this." and "What are the long term consequences?"

Although I had what is probably one of the shortest careers as a landscape contractor, I believe I learned more in just a few short months than a lot of contractors learn in years. And the biggest lesson I learned is if you are going to jump into deep water, you better have a life jacket even if you are the best swimmer in the world. Very few people in this country can afford to quit one job and work for a different company for almost two months without a substantial paycheck. At least not a person who has two sons, a month old baby, a car payment, house payment, and various other bills. Creditors don't care whether you get paid or not, they only care if they get paid.

The other important lesson which I learned is you really should pay close attention and know who you are hooking your wages to. On a personal level, Jimmy and I got along great. We very rarely argued (until there was no money), and we had a mutual respect for each other. He respected me because of the business knowledge I possessed, and I respected him because of the construction knowledge he had. You also need to make sure when and if you are going to do jobs for the government, MAKE SURE you have a decent amount of money in the bank to cover the higher wages required for that type of job.

And last but not least, the most important thing to keep in mind is - the grass in not always greener on the other side!

✦ I Was Involved In Family Politics

Martin began working in the family garden center in 1963 while attending high school. The family had been in business since 1955 and had an exceptional reputation. He worked part time while finishing high school as well as while attending the university. After receiving his degree in horticulture in 1972, he oversaw the operations of five garden centers.

Not one to pass up responsibility (or opportunity) Martin quickly began looking for ways to improve the operation of the garden centers as well as increasing profits. Product lines and costs were analyzed, operational efficiency was attacked, and in short order, the garden centers were set on a course of greater productivity and profits.

In the mid 1980's, the fortunes of the parent company began to take a turn for the worse. The parent company had grown quite large and diversified and it was decided to begin consolidating their operations. The company had decided it was in their best interest to discontinue the landscape business. Martin had been building this side of the business for many years and had taken a special interest in this area. He had seen the landscape division grow from a very low point, and felt that it would be a terrible mistake to shut it down.

Many discussions were held, but Martin was outnumbered. As can sometimes be the case in family-operated businesses, friction between family members began to mount. According to Martin, one of the biggest problems arose from the fact he was the only individual in the family who had any type of official schooling within the horticulture field. This led to major differences in business philosophies and signaled the beginning of the end.

For many years the friction continued, and it got worse as the time progressed. Martin tried to make his views known, but to no avail. His solution to the problem was "to bite his tongue and keep quiet." For three and a half years, he sat back and went along with the family desires. On many occasions he thought about quitting, but for one reason or another, he decided not to do so. However, the only reason that he survived the last three and a half years was he knew that he would be leaving the family operation eventually.

When the parent company finally decided to close the landscape division, it made Martin's decision very easy. Martin would start his own landscape business.

There were many things that worried Martin about starting his own business. He had only a small amount of money which was not enough to finance the starting of a business, but worse than this was the fact he had no other alternative. The situation within the family business had completely deteriorated. He had the choice of starting his own business, working for someone else (not the ideal choice) or being totally unemployed. Fortunately, his first (and only real) option worked for him. He began conversing with the local banks, and after many lengthy conversations, was able to convince the local banker he was worthy of a loan.

With the small amount of money he was loaned, Martin quit the family business and began his own landscape company. He put his budget into effect and hired people to work in the field as well as assist him in running his office. Although there have been many trying times in his new business, Martin says he was extremely happy.

He was asked to list what he felt were the most frightening aspects of starting the new business, and his answers were very clear. Although he had attended college and been working within the industry for many years, he was unsure as how to set up his budget. He knew the basics of what he needed, but had no real way of knowing if he had covered all of the bases. He had no financing at the time and was very doubtful as to whether he would be able to get any. The second most frightening thing was his lack of alternatives. As stated earlier, he had the choice of a) starting his own landscape business or b) working for someone else or c) being unemployed.

Martin has been in business for just over two years now, and things have been going well. He states that, "I still need to find my niche in the market place, but things are slowly settling down and we are starting to move at a steady, comfortable pace. We have had our share of ups and downs as with any new business, but we just take each item as it happens and work through it before tackling the next. One of the best parts of this whole ordeal is that the family business is still able to operate without me and the friction is gone."

Looking back over his many years in the Green Industry, Martin says that were a few up and down points throughout his career. His low points were not being given enough opportunities for advancement (family politics) and not having his views listened to seriously. His high point or what he considered the best thing to ever happen to him came at the point in his life when he decided that he had to face the choices in front of him. For four years he had been battling with himself and losing.

For four years he had suppressed his dreams and fears to just "go with flow," always afraid to upset the apple cart. But when the time came to finally make the break, he felt as if the world had been lifted from his shoulders. Of course he was now facing an even larger burden, starting his own business, but determination was the key. Determination, and the incentive of being able to prove all along he had been right, and he was capable of doing it himself.

Asked if he ever regrets what happened, he reluctantly answers with a "yes." He is not sorry he is in business for himself, and is not sorry for any of the many problems which he faces. His only regret is the extent to which the "family politics" became such a problem. Martin feels the problems should have been much easier to solve because it was a family owned and operated business and had been since it's inception. But as is evident by Martin's story, sometimes having a family business can be more detrimental than helpful.

✦ I Had To Get Out Of A Partnership

Bill grew up in a family of landscapers. When I say family, I mean family. Grandfather, uncles and cousins were all in it and it was the same company. It was a growing and thriving company in the sixties, but before Bill could rise to a position, his inheritance and future were sold for cents on the dollar. The company had failed.

But Bill was not to be deprived of his day in the sun. He started his own company. Before the family business failed, he and his young wife had taken the plunge and bought a brand new house. It was just after the closing on the new house that he attended a meeting that informed him of the problems within the company. To make matters worse, it had only been a short time since he had taken out another loan of $10,000 in order to buy stock in the now declining family business. He was now under pressure to make it on his own and make the kind of money that would get him out of the financial hole he found himself in now.

However, Bill found going it alone was a lonely situation, especially for someone who had always been surrounded by other family members. Bill, as he later would discover, was suffering with a problem of a low self-esteem which led to his being codependent on others. With all of this working against him, he began to search for a partner. He found one in what he thought would be a safe place to find a partner, at the church he attended. He had been doing construction, but his partner had a green thumb and so they started a maintenance division which began to take off.

Both men had limited business knowledge but Bill was good at production and his new partner was good at sales and like a lot of people in this business, they thought that if you could sell it and get it and get it done, you should make money. Make money they did, but manage and control it they did not and so the company continued to struggle.

As if one partner was not enough, they went out and got another who took on the new division of the interior plant maintenance. But within two years, it all began to fall apart. Bill's partners began to compete with each other as to who could spend most of the company money. They also began to be very jealous of each other. At the same time, Bill, who out of guilt had been working horrendous hours and whose self-esteem was even more shattered due to his relationships with the partners, was going through a divorce because his wife could no longer stand the problems. To compound problems, the newest partner who had only been there for four years, left the company to form his own company.

In a year, Bill remarried and his new wife believed in him and saw his potential. She came into the company to work in the office and help her new husband. However, what she found would disturb her very much. I will not describe all of the problems here but I will say that they were cause for dissolving the partnership. When she brought them to Bill's attention she found he had suspected these things for years but was afraid to even check the books or talk with people because he might have his suspicions verified. Now his wife knew the facts.

You see, Bill felt trapped. He lived in fear of his partner leaving and taking part of the business and leaving him alone to run the remaining part of the business. Because of that fear, he could not deal with the situation.

But there is good news! Bill found enough courage to face the problem. With the assistance of family, professionals and some trusted employees, he was able to work a deal which relieved him of that partnership.

It almost sounds like a soap-opera, doesn't it? I have said over and over that partnerships are like marriages. Some of the same dynamics work or don't work in a marriage will also affect a partnership. Anyone involved in a partnership needs to be ready to work at the relationship, and must also be ready to confront their partners at the slightest sign of difficulty in order to keep things from getting out of proportion.

When I asked Bill what he would do differently, he gave me three important items to consider for everyone who is thinking about a partnership. First, he would never give anyone part of the business just to have a partner. In the future, he will make any partners he takes on either pay for their share or earn it over a period of time. Anytime you give someone something, one of two things will happen: he will either have little respect for what was free, or will realize you have a great need which caused you to give them something for nothing. Either of those things will put you at a disadvantage.

Next, as already mentioned, he would address all the problems which surface immediately rather than let them fester. He finally feels he has come to a place where he can do this, and this fact now gives him the right to take on another partner. In fact, Bill says he feels a person going into a partnership must enjoy the kind of encounters which can make or break a relationship. They must not mind occasional shouting matches, but they must be able to walk away from those kinds of situations with something positive being resolved out of the encounter.

Third, he would never be as "naïve" as he was when he took on his first partners. Just as none of us would marry the first person we date after our first date, partnerships should be created only after a long dating and engagement period in which we get a chance to really see each other in the good times and the bad times. Some encounters should be experienced so you can see how you and your future partner work through some of these kinds of things.

However, Bill is glad for a few things which came out of his disastrous partnership and is even prepared mentally and emotionally to enter into another one in the future. He was glad for the partnership because it eventually forced him into standing up to the situation. Having done that has been a great boost to his self-esteem, and has caused him to get professional help so he can deal with his co-dependency. He feels he has also learned some good things from his partners who have helped him in his development. Surprisingly, he has learned from both the good attributes as well as the bad attributes of his previous partners.

However, he feels the best thing he ever did was to keep the company intact while dissolving of the partnership. This helped keep his good employees going and his customers satisfied with the least amount of trauma.

If Bill could say just one thing to all of you it would be to ask a simple question. If you are considering a partnership, why? If it is because of your own fears and low self-esteem or co-dependency, you may very well want to face what Bill has faced. If it is because you want someone to complement your existing strengths and make business better because there will be two equals who are different pulling the load, and you are willing to WORK at the relationship, then go for it.

✦ *I Bought A Company and Lost*

Mark, is a husband, father, and a mechanical engineer. He spent three years in the Air Force and nine years as a design engineer with a public utility company. At the age of 38 years old, he decided to follow the American Dream to own his own company. As he states, he did not want to go through life thinking he did not have the guts to go in business for himself.

With close to $100,000, he decided not to pull one up by the boot straps, but rather to buy an existing business of some kind. What he found was a contracting business doing four million dollars plus in sales run by a three person partnership.

Mark felt the business would be a good buy for several reasons. First, it had been managed by a partnership. The partnership was talented in their different areas of expertise. The talent had brought the company to its present level, but another person would be needed to take it to its next level. Mark felt he could be the person who would iron out some of the present lack of systems management which would allow the company to grow.

Second, the partners had all agreed as part of the sale to stay on with the company for a substantial period of time. Two of the partners would stay on as full time key employees; the third would stay a short while as a consultant. Mark felt this would compensate for his lack of knowledge of the business and give him a chance to learn the ropes before any of them left the company.

Third, he was an engineer with construction experience. He believed this combination would help him to manage a construction company.

During the first year after his purchase, things went great. The company showed a profit and seemed to be doing well. As we would later discover, this was a false indication of how things were really going. The company at the time of the buy-out was through the worst part of a big job. All the costs for that worst part were in the pre-buy-out figures. After the buyout, the good times hit on that big job which then began to show some good profit, making the first year after the buyout look good.

But then came the second year. All of the sudden the reality of the debt service began to set in upon the company. Along with that, several unsolved management problems began to surface and show their ugly heads. First, was the fact that under new management, sales had not increased. This was partly due to the fact the big job had put the sales staff to sleep and partly because there just wasn't work in the economy of this area at the time.

More importantly, Mark began to discover some tremendous personality conflicts between the previous partners. These conflicts had always been there, but now there was new management, they became an ever-increasing problem as people began to polarize to their different positions. Soon Mark found himself trying to tie together a team which was expending most of its emotional time and energy positioning themselves favorably at the expense of the company. The team was becoming more and more scattered in their approaches and positions. The previous partners who stayed on as employees gradually adopted a consistently adversarial relationship with their new employer and in fact attempted to run the business as if they were still the general managers, a situation that Mark, the engineer, was not trained to handle.

Mark began to find out what many people find out too late; it takes a mountain of cash to cash flow a construction business. More people have gone off the contracting scene because of a lack of cash flow than any other reason. The money Mark put into the business did not go into the business but into the partners. Thus, Mark was forced to keep a tight cash flow business going with increased overhead but no new influx of cash. It worked for the first year when the good job was going and was bringing in much needed cash, but it all came into reality when that job was done and he faced his second year without a "good" cash producing job.

After a little over two years of what became a personal nightmare, Marked walked away from the business by giving it back to the previous owners with little to no cash compensation for what he had put into the business. When I sat down to talk with Mark, he told me there are things he would have done differently if he had to do it all over again.

"I never will regret the fact I attempted to be in business for myself," Mark says. "However, there are certainly some things I would DO very differently.

"On top of the list, I would never buy into a big established business. I should have taken the money I had and started small, then let the company as well as me grow at the same time. If you are established in a business, it may be appropriate for you to buy a big business, but not if your first experience is with a big business."

"The other problem which occurred from buying into a big established business was I had to keep this big business going during the changes in leadership. I did not have the luxury of downsizing it until certain things could be worked out and then bring it back up to its previous sales level. Everything, debt service, overhead, etc., had been geared toward the company continuing on at its present sales level."

"Second, I would never buy a business with key people who I could not replace place of at any given time. Especially if those key people are the previous owners who, no matter what is said, still hold a powerful position in the eyes of the clients and the employees. I found the situation almost insurmountable as I tried to take the reins of the company. And when the personality problem surfaced, it really caused some nightmares as well as morale problems throughout the company."

"Finally, I did not listen to outside advice. This is not to say that I did not have outside advice, I just did not listen to it. I had a accountant look at the business, and he clearly advised me not to go any further. After he had looked at it, he could tell me several reasons (which I have already mentioned) why I should not buy the business. But I chose not to listen to that outside advice. I also made the mistake of keeping the same legal and accounting advisors that had been previous advisors to the company. This did not give me a complete and fresh perspective on what was going on within the company.

Today Mark, is working for someone else. Will he be back as a person who follow the American Dream and own his own business? I think so. And when he comes back, he will be like so many of us who go through a bad experience to take away our naiveté and then come back to have a MUCH greater chance of succeeding the second time around.

Taking Care of Business!
Are You Running It or
Is It Running You?

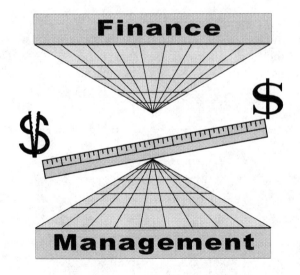

Lean and Mean or Fat and Sassy

"Fat and sassy" contractors are contractors who have become overconfident in their approach to this business. They came into the business during "the good times" and have been able to obtain work at fantastically high prices. This has allowed them to build up a dinosaur of a company to perform the work, and they still make a profit. They have a load of people on overhead to take care of every little thing that comes up—and more to cover anything that might happen. They have glutted themselves with highly paid management people and are charging customers for their inefficiencies.

"Lean and mean" contractors, on the other hand, base their entire approach to the industry upon the concept that they are not going to allow any fat on their management bones. Every employee working for the company must be productive and cost effective. Everyone must bring in, or help someone else to bring in, money to the company. Before hiring additional people, the existing staff accomplish what needs to be done until it becomes impossible or unreasonable. Any new hiring is done reluctantly, and only after careful consideration about whether such a hiring will keep the company lean and mean.

American business, as a whole, has been fat and sassy for many years. We have had a corner on the world market as well as on our own. This has caused unions, I believe, to sometimes demand unrealistic prices for their labor. It has caused companies to take on expensive people in management positions, who, if they were working for themselves, could neither justify nor produce the kind of dollars that they are taking as a salary. Plus, they have company jets, huge expense accounts, and expensive company cars.

However, we have seen the beginning of a new day in the American economy. Our products, made by fat and sassy companies, are losing their desirability here and abroad due to their high prices and poor quality. Imports are exceeding exports because of this very fact. Japan, Korea, and a half-dozen other countries, are whipping American business up one side and down the other. Why? I believe it is because managers in these countries think of quality over quantity, and put making money before making the Fortune 500 list. They are focused on being lean and mean rather than becoming fat and sassy.

What does all of this have to do with construction? I propose that it has a lot to do with construction! American business, including in the construction industry, is going through a much-needed change in how companies are directed and managed. We are being forced, by these world neighbors, into this very important rethinking of our approach to business as a whole. I predict in order to survive in the coming years, all American business will need to adopt a management style which is lean and mean.

I believe in this concept so strongly that one of the first things my consultants do when they go into a company is to look for the fat and then make suggestions as to how the company can become more lean and mean. That kind of management style puts more money on the bottom line and will protect the company from becoming a casualty as soon as hard times hit the economy in the company's part of the world.

What causes a company to become fat and sassy or lean and mean? I believe it all begins with the owner's, or manager's attitude. Attitude dictates the overall approach a person will take in administering and managing a company. I want to talk about some of the things which contribute to these approaches, as well as some guidelines to use in assessing whether you are lean and mean or fat and sassy.

A company may become fat and sassy when too many people are working in the office of a construction company. It is over administered. More things are being done than a company can afford to do. In the next section, I will go into more depth concerning this very problem. As you read it, understand that finding the balance between too much or too little administration will have a lot to do with staying lean and mean.

A second problem occurs when people are paid more than their job is worth. Notice I did not say more than they are worth, but more than their job is worth. That is an important distinction. One of the problems we often experience is the various personalities within our company. We make decisions about how much to pay someone, not based on what their job is worth to the company, but rather, on this person's need—in their estimation. This problem intensifies after a person is with you for a while and you develop a relationship with them and get to know their families.

I hate to give national averages, because people live and die by them. However, in this case I must make an exception and give you a national average that we have found to be a good rule of thumb. In construction, all overhead salaries should not exceed 8 to 15 percent of gross sales. The 8 percent is for those companies doing bigger jobs and having very little work subcontracted. Also, their work is mostly bid off someone else's drawings. It can get as high as 12 percent if you are having to design and sell your work, and/or your work is mostly smaller jobs with very little of the work being subcontracted.

Another problem I have found occurs when a manager views his or her business as a good place to write off things for tax purposes. There are tax advantages to owning a business, but that does not matter when the expenditures are unnecessary. If you don't really need something or someone, it doesn't matter if it is tax deductible. Look out! You may be in danger of becoming fat and sassy. If an expenditure cannot be justified as a good business decision on its own merit, don't buy it.

I would much rather pay taxes on money that I need for expansion, or as a safety valve between me and bankruptcy, than to waste it on something or someone that I don't need—just to keep from paying taxes. Now, not all expenditures that have a potential tax savings fit into the category which I have just presented. But before you spend money to keep from paying taxes, you had better evaluate it. Make sure the decision is based on more than just a tax savings.

How does one assess whether he is fat and sassy or lean and mean? Because these conditions are caused, for the most part, by one's attitudes, this can be very difficult. I believe the first step is to understand three things. First, because attitudes are often affected by our environment and feelings, we need to do whatever it takes to evaluate how environment and feelings—very changeable factors—are affecting our decisions as we manage our company. We need to review our company constantly, and how we are running it, to make sure that as time has gone by we have not allowed ourselves to become fat and sassy without realizing it.

Thirdly, we should not rely solely on our personal opinion when our company is lean and mean and not fat and sassy. I doubt if many of you reading this book right now would consider your company to be fat and sassy in its management approach. Yet, a good many of yours are! Every company needs somebody to be looking in from the outside. This will help them see things about themselves that they cannot see about their own company. I would suggest you get a good business consultant, or at least discuss this concept with a mentor who can advise you from an objective point-of-view.

Here is a checklist of points for you to consider.

- Could I hire someone to do my job for less money than I pay myself?

- Could I hire people for less money than I am paying my people? Are there more qualified people in the labor pool of my area who are readily available?

- Does the company actually need everything done in the time it is currently being done? Do we even need all those things done? Do I need all of the reports I think I do? Are there more inexpensive ways of getting these things done?

- What expenses are producing income, and which ones could I really do without? How can I, without looking like a cheapskate, encourage others to watch every expense?

I believe an owner or manager of a construction company must wear many different hats. He could be compared to a juggler, constantly juggling several balls in the air at one time. The balls are good estimating strategy, good field performance, good administration, and vigilance in watching where every dollar goes.

Part of the juggling act is to know when to allow certain expenditures and hiring, and when not to allow them. It is knowing how to constantly keep everyone's attitude in tune with expenses and profit. It is knowing how to sell, without being overbearing, the attitude says, "We as a business must remain lean and mean or we will be drowned in our own fat." Remember this: No one else in your company will even begin to consider these things unless you communicate them to your people. Even then, they will not think about them, unless you communicate them in a healthy and motivating way.

Let's talk about finding the fine line in how much administration you need.

How Much Administration is Too Much?
How Much Is Not Enough?

One of the most difficult things to establish is how much administration my company should be performing. How much is too much and how much is not enough? How many systems should I have in place? How many reports should I require, and on what areas of my business? Am I receiving enough information to effectively run my business? Or, am I not receiving something that I should, resulting in ignorance in a certain area, which will eventually cost me dearly?

I remember talking once with some management people from a certain construction company. The owner was an engineer who had received his degree from a respected university. Immediately upon graduation, he started his own construction company. At the time I was talking with him, he was working seven days a week, and in excess of twelve hours per day. He was paying himself a meager salary, and his company was not making any money. I soon discovered why.

Since he was an engineer, he was a very precise person who wanted to know every detail of what was happening in his company. He was, by national standards, a small-to-medium-sized contractor. Yet, as I listened to all of the systems he had in place to supply him with all the information he demanded, I realized he was doing things that only large contractors could afford to do. And, that was exactly the reason why he was working so many hours and not making any money both personally and as a company.

I must say, in my travels around the country, I have found this type of person is an exception. Most contractors are on the other end of the spectrum, and do not attempt to get nearly enough information to run their business. Toward the end of this section, I will discuss some of the minimums and maximums, as I see them.

You see, I believe there is a fine line that each company must locate between how much is too much and how much is not enough administration for them as a company. That fine line is a balance between what is necessary versus what can be afforded. To stay short of this fine line is to not get enough information. This is the more common error because, in the short term, it saves money. However, in the long term, it costs a great deal of money, because it causes contractors to make poor and uninformed business decisions.

To cross over that line is to do what our engineer friend did; to spend more money for information than you can afford given the current size of your company. This is often done because these things are seen as the right things to do—and they are. However, just because something is right does not mean you can always afford to have it. Other times it is done—and this is the worst reason—because the owner has seen or talked with a larger company who is doing it, and he wants it done in his company because the other company does it. I have seen this ego trip cost people their company.

The key is to find the fine line for your company. Once you have found it, go as close to it as you can without going over it. And, as I have said over and over again, recognize nothing stays the same in this business for very long and, therefore, this line will be moving constantly. You must remain flexible, and always be aware of where this line is in your company at any given moment.

Another principle that I wish to point out is that the construction business is always full of what I call "peaks and valleys." The following diagram graphically expresses what I am talking about. This first peak is what I call a Ma and Pa operation. I have found a contractor must do at least $250,000 worth of gross sales of work which is not subcontracted in order to make a decent living. It is difficult to operate much of a business doing any less. Ma and Pa can operate this kind of business themselves with Pa doing the estimating and running the jobs and Ma answering the phone, doing the book work and letter writing. I have found they can go as high as $350,000 doing it this way before estimating will get too difficult.

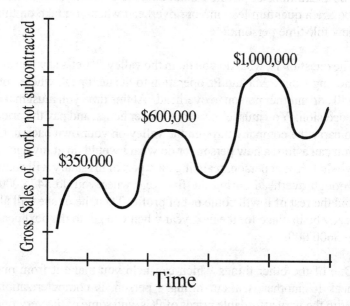

Peaks and valleys associated with growth of a contracting company.

I must qualify the dollar figures I have used in my examples with this explanation. The figures are based on a contractor who does most of the work with his own forces. Also, the average job is in the $10,000 range. If the jobs are larger or smaller, on the average, the figures used in the example will change, up or down.

There is a valley between $350,000 and the next peak of $550,000 to $650,000. Then there is another valley between $650,000 and $900,000 to $1,000,000. On and on goes the peaks and valleys, with the peaks getting farther apart the larger you become. When you are at the peak points, you are operating at maximum efficiency with the people you have in overhead positions. Each peak position usually requires the addition of a person to your overhead, as well as some associated overhead costs. These additions of people cannot be done gradually, but usually require a quantum leap in your overhead when you take on this new full-time person.

The question is, what do you do in the valleys? Let's say you are moving from a Ma and Pa operation to the next peak where you will add another person to overhead. At this time you must make a decision. You must either work longer hours, and pay the price to carry the company through the valley on your own back until you can afford a new person, or do what I would do at my age. I would hire that person, realizing only part of his salary will come through overhead during the first year when you do $475,000, and the rest of it will come out of profit. Then, he or she will already be in place for the next year when you get to the next peak at $600,000.

One of the other things which can help you make it from one peak to another, without hiring a person, is computerization. With the very affordable prices of PCs, and some of the very fine tried-and-tested software for this business, companies as small as Ma and Pa operations, who want to go from peak to peak, can do so with the aid of computers. The key to such a move is to get both your hardware and software from a consulting company that specializes in your type of business. I have seen many contractors sold a bill of goods because they hooked up with the wrong kind of person to supply their computer needs.

How much is too much and how much is not enough? Let me give you some quick guidelines which must be tempered with the understanding you are all so different, except for a few of you, these guidelines will not fit.

Every company, regardless of size, should have a monthly financial statement. If you are a smaller company, it can be done outside your office with a simple compilation of billings and of checks written from your checkbook. Larger companies should get a more sophisticated statement, breaking down income and expenses into divisions (or profit centers), if they are doing more than one kind of work. Again, I have found very few small companies that can afford this last type of accounting, even though they may be involved in many different things.

If you are a smaller company, payroll can be done by outside companies. It can also be extended to biweekly if you need to, as it costs money every time you put out a payroll, regardless of how many people you have on the payroll. Larger companies can afford to do payroll themselves, and to pay more often.

Job costing is one of the major areas to consider. Again, I believe every contractor, regardless of the size, should do job costs. Smaller companies should not spend the money to job cost every penny on every job. Rather, I recommend they just have a simple job costing system which tracks the most important and most risky part of construction: labor. I would recommend you simply keep track of the number of hours estimated versus the number of hours actually used. Larger companies should keep track of equipment, as well as material, through their job costing system.

Another major area concerns secretarial and phone answering help. For some smaller contractors, the best they can do is to use a phone answering machine, themselves, or their spouse. I find that, for most contractors, their first hired person (past the Ma and Pa stage) should be a versatile secretary who can also do light bookkeeping. The next major person hired is either an estimator/sales person or a person who can manage the field. Again, the decision of which one to hire will depend on which one most makes the owner better off than he would be otherwise.

I conclude by reminding you that there are no hard and fast rules in this area. Every company, and its approach to business, can be so different. The people who run it, the people they hire, and the amount of work they can do, vary greatly from one company to another. As I have said in previous sections, the key is to evaluate, evaluate, evaluate every chance you get. Administration—determining how much is enough and how much is too much takes a mixture of good management science and reading your gut feelings.

The Management Process

Management is a dynamic process, by which we accomplish goals and projects by ourselves or with other people. It is a process which involves four interrelated phases. A project begins as an idea and ends with a finished product which must then be managed and maintained. This principle is true, whether what is to be constructed is a building with landscape, a bridge, a ship, a book, a movie, a championship NFL football team, a successful construction business, or a successful family.

Management is not static. It flows from the design phase to the development stage to the management stage to the maintenance stage. One can easily see when a construction project goes through these four stages. First, it is just an idea, usually designed by an architect, in the conceptual or design stage. Then it enters the developmental or the build stage, when it is put out to bid, and a contract is signed with a contractor. Once the project is built, it must then be managed, or rented to tenants. Finally, it must be maintained as a finished project on an everyday basis. The same phases occur in the administration of your company. Every new venture, idea, or method of operating, must go through these same four stages. Also, you as the owner must personally carry it through these stages, or be able to hire and delegate the performance of these different functions.

Interestingly, people tend to gravitate toward one of these four categories. Some are gifted as designers, or idea creators who deal best with abstract and conceptual things. These include architects, designers, musicians, poets, authors, entrepreneurs, explorers and inventor-type people. Next are the developers who are gifted at figuring out how to make the idea happen, and then seeing to it that it does indeed happen. These include contractors, directors, and engineer-type people. They take the ideas (i.e. the designs, blueprints, manifestos, causes, business plans, etc.) of the designers and bring them into the reality of a finished product.

Managers are the ones who are best at managing one of the finished products. These include property managers, office managers, sports team managers, political campaign managers. Managers then rely on another group of people to properly manage that which has been handed down to them by the designer and the developer-type people. The final group include the maintenance-type people of the world. They are the ones who, on a daily basis, keep things running smoothly. They are the janitors, mechanics, lawn mowers, file clerks, and librarian-type people. They put the spit and polish on the brass as they keep things running.

Some people have the characteristics of two of these groups. For instance, some individuals are design/developer types. These may include contractors who cannot just be an architect, because they enjoy the building part of the business, so they become design/build contractors. There are also developer/manager types, who do not like design work, but involve themselves in bidding on pre-designed work, and then sometimes own the project themselves. Then there are the manager/maintainer types who like to manage a company of property, as well as maintain it.

You will notice that these combinations encompass two connected phases of the process. Rarely, if ever, will a designer enjoy the management or maintenance positions. Nor will a developer thrive in a maintenance environment, or a maintainer be creative in the designer position.

The key, as I pointed out in several places already, is that you know yourself, and your people, in light of this dynamic that I have just given you. Never make the mistake of placing a developer in a position requiring a manager or a maintainer (at least not for very long) and never put a designer in a maintenance position, and vice versa. Identify your people and your strengths and weaknesses in the light of what we have just discussed and then deploy your people using their or your strengths.

If you are a smaller company, you still need to understand this concept. Identify your gifts in the phases we have just described. It will not change the fact that you will still be weak in some areas (that are needed) but now you will at least know two things: first, the type of person who will be the first overhead person you should hire, and second, the need to be as creative as possible in your weak areas, finding help from someone outside the company, part time employees, or possibly by just leaning down and doing it yourself. If you choose the latter, at least you will understand why you are having a problem with this area. You should then try not to be too hard on yourself when you do not do quite as well in this position.

Once we understand the different phases of the management process and the various types of persons who gravitate toward them, we can implement a couple of tools which can help us to facilitate, as well as quicken, the process. This process will take a project (or idea) from the conceptual phase through the maintained finished product.

The first tool deals with projects/things you need to address on a daily basis. It also helps you to gain control of all other things which are bombarding you right now. I call it the *Project/Priority Worksheet* (see illustration on page 170). Use a standard yellow pad, and line it and label it just as I have. Then start entering in column 3 all of the things which are occupying your mind. Put down all of the loose ends, the frustrations, and any other items or things which are on your mind, that you feel are needing your attention at some time or another. Keep the descriptions of the items very short, and add any amplifying remarks in column 4. Allow a couple of blank lines between projects which have additional details to consider.

Next, identify your priorities (in pencil) by placing a "T" (for top priority) in column 2, next to the ones which require immediate attention. Then place an "L" in column 2 next to those things which you almost do not care about (yet). Identifying these two extremes should provide you with a good deal of latitude to determine the next two classifications. Identify the high priority items (the ones which do not yet have you to the point of pacing the floor) with an "H," and the medium priority items with an "M."

Our goal at this level is to merely identify the various items, spreading them out, in order to get a better perspective, and to determine which ones actually do need our attention, and which ones we need to ignore until they become top priority. Most of us are fairly good at what we do, but we often do the less important things at the expense of the important ones. There are some of us, however, who would also be good at doing things, but we don't accomplish what we should because we are too frustrated at all the things that are bombarding us on a daily basis.

Once you have penciled in your priorities, as you see them at this moment, begin attacking the items marked with a "T." You will now be handling each task in the proper order. Once you complete a particular item, or do all that is needed for now, put a line through it, or change the priority accordingly. Just seeing certain items crossed out, and knowing you have a plan, can be quite motivating to you, and can encourage you to be what you are doing right. To continue, simply keep going through the list, re-marking it as often as necessary (which could be as often as daily). Let me mention, however, a personal observation. I have never been able to get all of my T's completed. This is because the ones marked "H" "M," and "L," all become a "T" at some time or another. The key is to work on the right project at the right time. Do not get caught straightening the picture on the wall of a sinking ship.

If you are fortunate enough to have people working for you , you can use this tool to help you manage them. Write the name of the person to whom you delegate a project/item in pencil in column 1. I do not usually put in due dates on this worksheet but, if I need to, I place the due date under the priority which I have assigned it in column 2.

The project/priority work sheet is not overly complex. You can carry the pad of paper with you wherever you go.

The next tool is the *New Ideas/Projects Board* (see the illustration on page 225). It is designed to help you identify the process of capturing and storing new ideas, and then of implementing them in a timely manner. Use a 3' by 4' dry ink erasable white board like the bid board described earlier. Mark it just like the illustration. Use black markers only on it (later we will be very selective and identify really hot items with a red marker) or else you will hinder the ability of the board to focus your efforts and to motivate you.

The left side of the board helps you to capture new ideas. As new ideas occur to you, write them in column 3, and place a tick (-) beside each new idea to help distinguish them from the sub ideas, which are indented and written right below the main idea. Use the priority and delegation columns just as you would on the project/priority worksheet.

The projects side of the board is used exactly like the project/priority worksheet, only it is for longer projects. These are projects that will be actively in progress for the next 30 days to 12 months, or even longer. You simply want a place to put these types of projects that are impractical to deal with on the daily project/priority worksheet.

This board is designed to help you focus your management energy on crucial long term projects, and on new ideas which will greatly enhance the business. So often these are the very things many of us forget because we get so caught up in the everyday problems of running our business. This causes our businesses to stagnate, and even lose money, because we are not remaining on the cutting edge of all possible new directions our business may take.

There will probably be four or five items on the board which are the really hot projects or the really great ideas. These are the ones you will want to focus on and attack. Mark these with a small red star beside the particular priority item. Once you have identified the red or hot items, work on them until they become a reality. Draw a line through them when they are completed to help keep you motivated with the thought that you are completing some of these long range plans (that will set the direction for your company over the long haul).

Another tool that I want to discuss is the *Office Checklist* (OCL) (see illustrations on page 226 and 227). The OCL is a tool which helps you to monitor and control the routine and repetitive things which need to be accomplished. It is designed to allow you to quickly ensure all of these details are being handled. It also sets goals for individuals. It not only identifies who does these items, but it also identifies when they are due, and to who they should be submitted.

The OCL is simply a checklist (similar to a pilot's checklist that is used before taking off and landing) to ensure you have not forgotten anything. It does not explain how to do things. It should cover 75 to 80 percent of the items which are accomplished in your office. Use this format to help you create your own form on your computer.

The final tool I want to discuss is simply called *Weekly Schedule* (see illustration on page 228). It is a picture of the weekly and monthly portion of the OCL. Type the eight to ten major items which occur each week and when you want them accomplished. Keep the information brief. One or two word descriptions of each item are best. Do the same on the illustration for the monthly portion of the OCL.

I recommend that key people in your office have copies of the OCL, and each person develop their own picture schedules of the portion of the OCL for which each are responsible. They should have it displayed on or near their desk, in a place where they will naturally look during the course of their day.

In conclusion, I want to emphasize that management (accomplishing goals with the help of people) is not a mysterious process. It requires goals, people, and support systems for accomplishing these goals. The tools discussed above are simply that, tools. They can never replace sound judgment or, as I discussed earlier, the gut feeling that comes with experience. Remember, those two things (judgment and gut feeling) cannot be delegated. Do not think your use of these tools will magically bring instant results, in and of themselves. You will still need good common sense, and a drive to see things accomplished in an orderly fashion.

These tools concentrate on accomplishing goals, and on identifying who does what. They are not time-management tools. They are goal/project tools. After all, I believe that time-management simply does not exist. None of us can save it, spend it, invest it, or even waste it. Time is a medium through which you pursue your goals. Just as the captain of a ship pursues his goals (or destinations) through water, you and I pursue our goals through time. You are no more the manager of time than the captain is a manager of water. Concentrating on time-management is like concentrating on water management. It will distract you from the real issue at hand, accomplishing your goals. Clearly define your goals, and reduce them to bite-size, measurable chunks. Then pursue them with the help of the tools discussed above, and with the best use of your time. If the tools do not work for you, change them until they do.

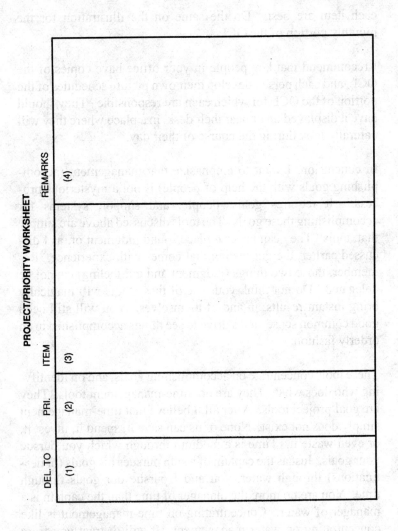

PROJECT/PRIORITY WORKSHEET

DEL. TO	PRI.	ITEM	REMARKS
(1)	(2)	(3)	(4)

Project/Priority Worksheet

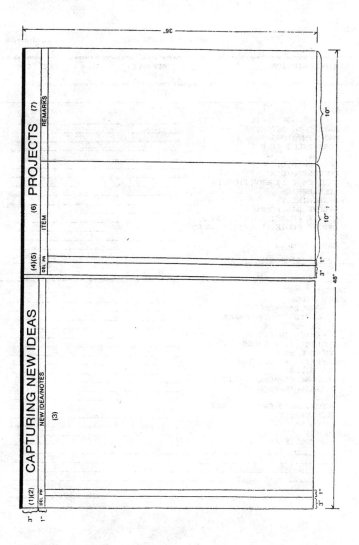

New Ideas/Projects Board

```
===================================================================
ITEM/DESCRIPTION                     PREPARED      TO     DUE  BY
                                     BY
===================================================================
                               DAILY
-------------------------------------------------------------------
(TASKS)
  PICK-UP/SORT MAIL               -----------  ----------- ---------
  MAKE BANK DEPOSITS              -----------  ----------- ---------
  MAIL INVOICES                   -----------  ----------- ---------
  POST BILLS TO P.C.              -----------  ----------- ---------
  FILING                          -----------  ----------- ---------
  BACK-UP P.C.                    -----------  ----------- ---------
  PROCESS P/O'S                   -----------  ----------- ---------
  PROCESS DELIVERY TICKETS        -----------  ----------- ---------
  LOCK OFFICE                     -----------  ----------- ---------
(CHARTS/MISC.)                    -----------  ----------- ---------
  UPDATE BID BOARD                -----------  ----------- ---------
  UPDATE EQUIP. MAINT. BOARD      -----------  ----------- ---------
  UPDATE EQUIP. LOCATION BOARD    -----------  ----------- ---------
(REPORTS)                         -----------  ----------- ---------
  CASH ON HAND                    -----------  ----------- ---------
-----------------------------     -----------  ----------- ---------
                               WEEKLY
-------------------------------------------------------------------
(TASKS)
  NEW JOBS:                       -----------  ----------- ---------
    -PRELIMS/RELEASES             -----------  ----------- ---------
    -CERTS. OF INSURANCE          -----------  ----------- ---------
    -SET-UP FILE                  -----------  ----------- ---------
    -ENTER IN P.C.                -----------  ----------- ---------
  COLLECT TIME CARDS              -----------  ----------- ---------
  APPROVE BILLS                   -----------  ----------- ---------
  DEPOSIT PAYROLL TAXES           -----------  ----------- ---------
  PAY BILLS                       -----------  ----------- ---------
  PREPARE PAYROLL CHECKS          -----------  ----------- ---------
  DISTRIBUTE PAYCHECKS            -----------  ----------- ---------
(CHARTS)                          -----------  ----------- ---------
  RECEIVABLES/PAYABLES RATIO      -----------  ----------- ---------
  RECEIVABLES                     -----------  ----------- ---------
  PAYABLES                        -----------  ----------- ---------
(REPORTS)                         -----------  ----------- ---------
  JOB-COST EACH JOB               -----------  ----------- ---------
-----------------------------     -----------  ----------- ---------
-----------------------------     -----------  ----------- ---------
-----------------------------     -----------  ----------- ---------
-----------------------------     -----------  ----------- ---------
-----------------------------     -----------  ----------- ---------
-----------------------------     -----------  ----------- ---------
```

Office Check List - Part 1

```
                              MONTHLY
------------------------------------------------------------------------
(TASKS)
   FINANCIAL DATA TO CPA          ----------  ----------  ---------
   RECONCILE BANK STATEMENT       ----------  ----------  ---------
   --------------------------     ----------  ----------  ---------
   --------------------------     ----------  ----------  ---------
   --------------------------     ----------  ----------  ---------
(CHARTS)                          ----------  ----------  ---------
   MATERIAL/LABOR RATIO           ----------  ----------  ---------
(REPORTS)
   FINANCIAL STATEMENTS FROM CPA  ----------  ----------  ---------
   CASHFLOW BALANCE               ----------  ----------  ---------
   BUDGET VS. ACTUAL              ----------  ----------  ---------
   --------------------------     ----------  ----------  ---------
   --------------------------     ----------  ----------  ---------
------------------------------------------------------------------------
                              QUARTERLY
------------------------------------------------------------------------
(TASKS)
   PAYROLL ADJUSTMENTS FOR INSUR. ----------  ----------  ---------
   --------------------------     ----------  ----------  ---------
   --------------------------     ----------  ----------  ---------
   --------------------------     ----------  ----------  ---------
   --------------------------     ----------  ----------  ---------
------------------------------------------------------------------------
                              YEARLY
------------------------------------------------------------------------
(TASKS)
   PREPARE F.Y. END DATA          ----------  ----------  ---------
   PREPARE F.Y. CLOSE-OUT         ----------  ----------  ---------
   REVIEW CORP. TAX PICTURE       ----------  ----------  ---------
   FILE CORP. TAXES               ----------  ----------  ---------
   PREPARE/DISTRIBUTE W-2'S       ----------  ----------  ---------
   PREPARE/DISTRIBUTE 1099'S      ----------  ----------  ---------
   REVIEW INSURANCE (GL & EQUIP)  ----------  ----------  ---------
   RENEW INSURANCE                ----------  ----------  ---------
   REVIEW MED. INSURANCE          ----------  ----------  ---------
   RENEW MED. INSURANCE           ----------  ----------  ---------
   REVIEW W.C. INSURANCE          ----------  ----------  ---------
   RENEW W.C. INSURANCE           ----------  ----------  ---------
   OBTAIN W.C. DIVIDEND           ----------  ----------  ---------
   OBTAIN W.C. EXPERIENCE MOD.    ----------  ----------  ---------
   START BUDGET REVIEW            ----------  ----------  ---------
   BUDGET IN P.C.                 ----------  ----------  ---------
(REPORTS)
   FISCAL YEAR END (F.Y.E.)       ----------  ----------  ---------
   --------------------------     ----------  ----------  ---------
   --------------------------     ----------  ----------  ---------
   --------------------------     ----------  ----------  ---------
```

Office Check List - Part 2

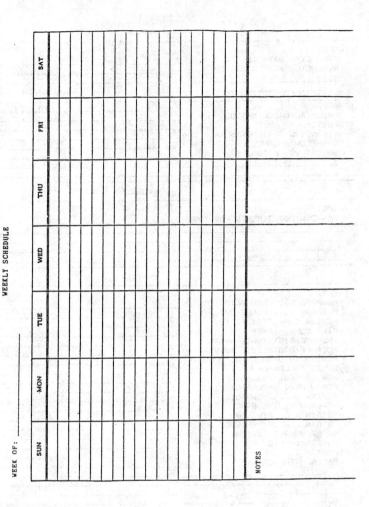

Weekly Schedule Form

Storing Information

It is important, in this section, to deal with what may seem to be a very simple situation, yet, in reality is one which can become a very big problem for many contractors. If this item is not handled properly, it can slow an office down to a crawl, and waste many hours and dollars, as well as cause a contractor and their people great frustration. Let's talk about storing information in such a way when is easily retrieved by those who need it, when they need it.

There are companies whose office always looks as if a tornado had just passed through. Items are often stored in boxes. Papers are on shelves, on top of file cabinets (instead of in them), and stuffed into the drawers of desks. Once or twice a year, for a week at a time, everything is examined and either thrown away or handled. Sometimes, jobs that are five months old are discovered to have never been billed. Jobs are also discovered which had been contracted, but never done! When the people are called, they have already given the job to someone else after they realized they had been forgotten. Thousands of dollars are lost to this kind of paper-management. To help one in this area, here are some very simple rules to keep in mind as you work with information (such as papers, brochures, documents, publications, magazines, reports, computers diskettes, etc.):

1. Everything should have a "home" or a least be heading toward its home. Sit down, think through, and lay out a logical system that meets your needs and encompasses all of the items listed above which need to be stored.

2. Keep it simple. Remember that the purpose of a file system is simply to enable a contractor and his employees to quickly and effectively store, find, and

then to apply, information. Do not design a system which is usable by only one person in an office. When you are designing the system, elicit input from everyone who will use it.

3. Keep it all encompassing. Identify the different items which must be filed, and draw a rough diagram indicating where they will go. Be certain to not only include the final destination, but also where they will be in route to that destination (e.g., field baskets, incoming box, pending box, outgoing box, etc.).

4. Keep file cabinets categorical. Identify the major categories to be included in a file cabinet. Such categories could include insurance, finance, equipment, licenses, legal, training, personnel, marketing, operations etc. Once the major categories have been established, identify the sub-categories within each major category. Under insurance one might have health/medical, general liability, Workman's Compensation, etc. Finance could be a huge category since just about anything can be tied to finance in one way or another. In this category, only include items directly involved in financial matters. These would include, but would not be limited to, such things as financial statements, line of credit information, tax related items, loans, financial budget projections, bank account information, etc.

The rule is: file an item in the category in which one will most logically look when they want to retrieve it. In some cases, one might want to file something in more than one location. Copy it, then file it in several locations. For instance, you might receive an estimate of damages on one of your vehicles which was involved in an accident. You would probably not file it under finance but, rather, under insurance, and then possibly under a particular piece of equipment, as a sub-category of "equipment."

5. File items quickly. File them before they can get lost! At a minimum, file the important items quickly, and put the remainder in a "to-be-filed" basket. Do not, however, let that basket sit with anything in it for more than just a few days.

6. Keep it flexible and expandable. Always keep on hand plenty of cabinet space for sorting file items, and be prepared to quickly add a new category, or sub-category, when needed. Often contractors attempt to save $100 for a cabinet and, thus, throw their filing system into a shamble. Also, keep a lot of extra file folders and baskets available, even if they are only needed for a short period of time. Get things into a "home" as quickly as possible rather than have them floating around.

7. Keep it visible. Do not allow items that need to be filed to be stuffed into desk drawers where they may be out of sight. At that point, they are also lost. Keep everything in a file flow system.

8. Let certain employees have their own informational and personal files. These are very important, as some employees often need to have quick access to information to do their jobs. Help them to do so by having file cabinets, and other personal resources, available for them. However, ensure the important information which the company needs, is kept in the formal company files. Also, if and when an employee leaves the company, be sure that his or her files stay with the company.

9. The waste basket may be one of the most important files. It is important one knows when to "hold-em and when to "trash-em." Do not keep an item unless you feel that you may use it in the future. Be careful to not save things solely out of fear when sometime, maybe, you may "possibly" want to see it again.

10. Magazines and copied articles from publications should be stored in file boxes which can be purchased from a local office supply store. Books should be stored on a shelf, or bookcase, in some form of alphabetical order (or subject order) so anyone can easily locate them when they need them.

Legal size file cabinets are preferable because of the flexibility and expandability that they offer. When you are putting headings on files, pencil them in. This allows you to add files or to quickly erase and change headings if they become obsolete, or if you discover a better way to label the file.

Tri-cut file folders are preferable. Someone can then put all main categories on the left tabbed folders, and sub-categories on the middle and right tabbed folders. This will allow employees to be able to search the left row with one eye movement for a major category. Once they have found that, they can move over to find the sub-categories. Sub-categories in the same rows will confuse a file system and will waste time when someone needs to locate a file quickly.

Management Styles

In the last section we dealt with a very nuts-and-bolts issue. Now, I want to return to some theoretical issues. Earlier in this book, I dedicated a section to management by team work (page 171). In that section, I share from my own experience of that style of management. I want to take some space here to expound on two other forms of management by team work which may differ from the particular style I explained earlier.

First, regardless of which style of management you adopt, every person within your plan should have a detailed job description. Even though job descriptions have been a standard management tool for years, it is amazing how many contractors do not use them, though they may have heard of them. Even if you are a Ma and Pa operation, Ma and Pa should each have a job description. This will clarify who is doing what, and everything is being handled by someone.

For those of you who have never written one, let me go over a few steps you should take in writing one. First, list everything that needs to be done, and administered, in the entire company. Then begin to place names by each item as you determine who should take care of each one. When you do this, you will take into consideration the management style you have chosen from the ones we will describe in this section. Also, you will take into consideration the various types of people we discussed earlier.

Then, take this rough draft, and sit down with each individual to get their response, not about what you have assigned to others, but on what you have assigned to them. Continue to work with the main list until you, and the others involved, feel fairly comfortable with it. I say fairly: you will almost never achieve total agreement on this list.

From this list you are now ready to write individual job descriptions for each individual. These job descriptions should be as inclusive as possible, without inhibiting people from being creative, or from doing things outside their job description. However, you need to emphasize to each person that they need to fulfill their job description first before taking on new or different projects. These descriptions should be as simple as possible. They are just an outline, or listing, of the things a person is responsible to perform.

Everyone should understand that job descriptions are not written in stone. As new tasks or unforeseen things develop, someone may be asked to perform something which is not on his or her job description. Also, every person's job description should be reviewed—and possibly changed from time to time—at least every six months.

Before I explain the two management styles, let me reiterate a major principle involved in management. I believe construction is one of the most difficult businesses to run. In all of my travels around the country, meeting thousands of contractors, I have only met two of them who had a master's degree in business. Students spend six or seven years in college, studying every kind of business there is, then, when they finally earn their degrees, guess which business they *don't* go into. This one. After comparing construction to other businesses, they go into almost any other. Does this tell you something about our business?

The construction business is very difficult to manage because it is filled with so many variables. You have variable jobs with variable site conditions, and you work for a variety of owners and architects. You have a variable labor force, variable weather, and a variable overhead, which you recover off variable job costs! Don't fool yourself. This can be a very difficult business to run. But it is not impossible.

Run your construction company with as many eyes, ears, and minds as possible. That's why I believe you should form a management team to help run your company under one of the following two management styles.

I operated under the first style for many years. It included three key people who formed the team for the entire company and for all its divisions, or profit centers. We were each responsible to the owner, and to each other, and handled everything within our own spheres, including various company divisions.

The three people were a contract administrator, an accountant, and a field superintendent. (I have described them in detail in the section on page 171.) As contract administrator, I estimated all jobs, negotiated them if it was necessary, and then brought them to contract. I then saw that all purchase orders and subcontracts were written, and a form of billing was established with the owner. I then conducted an interoffice preconstruction meeting, in order to cover all of the necessary details, before turning it over to the field superintendent.

I would remain involved with each job, pricing all change orders, doing the monthly billing with the field superintendent and accountant, and doing the punch list at the end of the project. I also looked over all the job-costing information on the job, compared it to my original estimate, and approved all invoices and subcontract payments. I attended any meetings which were necessary when problems came up on a job. As the company grew, I had several contract administrators working under me in this department.

After the preconstruction meeting with the owner, the field superintendent took the job and coordinated all the necessary men, materials, and equipment, and saw to the completion of the job. It was his responsibility to work with all of the field foremen. He made sure their jobs used only the estimated amount of labor. He hired and fired as needed. He also supervised the mechanics in the garage or sent equipment out for repairs, so it would be available when it was needed.

The accountant performed all of the accounting functions, as well as the role of office manager. She directed all office personnel. She also took care of paperwork, filing, and reports.

Style number two uses the same type of individuals, but places them in a different relationship to one another, and to the owner. A team of managers takes on the divisions, or profit centers, of the company. Possible divisions might include construction, maintenance design and build, equipment, and the accounting department. Each of these divisions has a manager in charge of everything in a division. They are then responsible to the owner and to each other.

A division manager may be a contract administrator, a field superintendent, or an accountant in the case of the accounting department. The decision as to who will be the manager does not have to do with job titles, but with who is the most qualified to run a particular division.

Once the owner establishes the team of managers, each manager creates a team of other people to run his division. One division will do for secretarial and accounting services, if the needs for each are not large enough to require separate divisions.

Consider the personalities, strengths, weaknesses, and goals of your people as you decide which way to run your company. You may have a preference you'd like to go with, but you should consider everything before you make a decision.

For example, I operated under the first style because it fit my personality and that of the field superintendent of the company. I believe to this day that neither one of us would have functioned very well under the second style of management. The first style just fit our personalities better.

However, I think the second style is much easier to run. It defines accountability more clearly, and it does not require perfect cooperation between the contract administrator and field superintendent. The first style does demand this kind of cooperation.

Either style requires you hold one major annual meeting. Hold it away from the office in a conference facility or resort. If you operated under the first style, the three key members of the management team should attend. If you operated under the second style, the division managers should attend.

The purpose of the annual meeting is the following:

1. To establish a company budget, which will include projected sales for each division, as well as projected costs and profit. These budgets should be reviewed by everyone in the meeting.

2. To brainstorm about future direction and goals for the company as a whole. This is the time to add items to your New Ideas/Projects Board (page 225). It is also a time to congratulate your people on the good job they have been doing, and show them the great potential which exists for your company. Guest speakers are not always a bad idea. I have spoken at some of these meetings.

3. To network among your people. This is a great time to let each of them share some of the problems they are experiencing in their division, or on a particular job. Your employees will contribute some of the best answers to your employees' problems—and the company's problems.

Sometimes, your company will make a great deal of income generated as a result of these types of meetings. I encourage every company, even a Ma and Pa operation, to spend some time on one each year.

Now, I want to address a question that is probably on the minds of some of you. What do I do when people don't perform under either style as they should? There are four things I want you to consider if it happens to you. First, examine yourself, and your opinion, to determine if your perception is accurate. I have found sometimes my opinion about people who are not performing may be flawed, because I am not seeing all the hurdles being jumped, and the extenuating circumstances which may be hindering performance. Also, I may be failing to totally evaluate the completed product produced by a person. You may need to get more information.

Second, reexamine the person's job description. It may be the performance is not what it should be because the person does not fully understand his or her job and what you expect. This can be a fault with the original description or simply due to the fact when something has changed, making the original description obsolete.

Third, reconsider the person's qualifications. You may have misread their ability or talent and have attempted to put a square peg in a round hole. Management is an endeavor that involves something of a trial-and-error process, because it involves people. It is never too late to make changes. Many times, I have found people more than glad to make such changes, because they recognize things are just not working.

Fourth, don't be afraid to let someone go. If someone does not fit in your company, and does not work well with your people, they will probably fit somewhere else, and there is someone who will fit better in your company. If you keep people where they do not belong, you do yourself—and the employee—a disservice.

I also want to point out an important principle concerning these management styles and people. I feel that, in most cases, it is best to get your administrative ducks in a row before you go out and acquire a lot of additional work. Upon this writing, I had just put down the telephone after talking with a contractor who testified to this. He first got his administrative functions in order, then went out and increased the sales of his company. There are too many contractors who get in a hurry to increase sales and get the cart before the horse. They increase sales without having the management people in place to handle it, then they suffer loss of income, and damage their reputation.

I conclude with a statement I often make to contractors. I tell them they need a full-time consultant to their company. When I say that, they often look at me with questioning eyes. I can tell they are calculating the cost of such a thing. They know they could never afford it. Others think I'm trying to drum up business for my consulting firm. I go on to explain to them the full-time consultant their company needs is *them*.

One of the greatest company owners and managers I have ever worked for had this very approach. He developed people like myself to run his company. Then, he let them do what he had trained them to do. However, he always told us when the going got tough and we needed advice, his door was always open, and that he was there to consult with us on any problem we faced. And, this is just how we viewed him, as our consultant. This style of management allowed us to gain from his experience and knowledge, without being intimidated. It allowed us—and the company—to grow to be all we could be.

Every one of you, no matter how large or small, has a management style—whether you want to or not. For some of you, your style has developed through the years with no conscious effort on your part. Others of you have followed some kind of plan. It does not matter which of these best describes you. It is important for you to do two things. Identify your style, and try to write it down or articulate it in another way so you and others can examine it. Then, with the help of others, constantly reevaluate your style and always attempt to improve it—make it fit you and your people.

Disciplining Your Money

Many contractors are in financial trouble, not because they are not making enough money, but because they are not managing their money properly. Here are some concepts about—and methods to use—to discipline money.

Be objective about your ability to handle money. Be realistic about your employees' or partners' abilities with money. There are people who simply cannot control their spending, let alone discipline the company's money. As soon as they see money in the check book—whether it be a personal checkbook or a company checkbook—they spend it.

Sound like you or an employee who has access to your company's coffers? Do them a favor. Set up a system of accountability that will keep you, the employee, and the company from being victimized. You can do this in several ways.

1. Turn the checkbook over to someone else so you or they must go to another person for a check. The "big spender" can still sign these checks, if necessary.

2. Have the accountant issue the payroll and monthly accounts-payable checks, and then set up another account from which the big spender can make minor and emergency expenditures. However, keep the balance in the second account as low as possible.

3. Set up a committee of levelheaded people who will advise the big spender about expenditures which exceed a specific amount.

Some of these things may sound embarrassing to a contractor; but they are less embarrassing than bankruptcy.

A second thing everyone needs to understand and do, is "down-and-dirty accounting." This simply means a contractor must acquire the ability to determine quickly where he stands financially. It is important to get monthly profit-and-loss statements. However, many times those statements don't reach a contractor in time for him or her to make decisions about spending money. Consequently, the contractor always needs to have a feel for where he stands financially. Many times, there may be money in the checkbook that looks like profit, but when you consider all of the upcoming bills, there is no profit for spending.

Contractors often receive large sums of money for their work. Rather than look at the bills which are coming due, some contractors look at the money they have in the checkbook, and do not apply any down-and-dirty accounting to that money. Consequently, they spend it—and find themselves in trouble down the road.

Many contractors keep approximate financial records in their own desk, just to maintain a daily "feel" as to where the company is financially. Accountants must submit an excellent bookkeeping record to them on a monthly basis, but the contractor keeps this personal record for the purpose of his own money discipline, as well.

The contractor should do this down-and-dirty accounting every month after having received a financial statement. Remember, just because a financial statement indicates that a contractor has made a profit, does not necessarily mean that he has made a profit. He must always question each statement's results until he is certain that everything has been included. If the statement does not show some outstanding payables or some unbilled jobs, the overall picture will be off. Every month, there should be an "over-and-under-billing" meeting. During this time reexamine every job and every account to make sure that the financial statement reflects reality.

Let's look at what can happen to a contractor. A financial statement shows a larger than normal profit, which is the first sign that something could be wrong. He begins to spend some of "his money." After a year, the insurance company audits the company and hands them a bill for almost $40,000 in additional insurance premiums for Workmen's Compensation and for liability on their payroll. Initially the insurance company quoted them—and they paid—a premium on an amount based on a payroll that has since doubled. They collected the additional money through their labor burden percentage, but it had not shown up as an expense in their financial statements. That was because the insurance company did not detect the increase until they audited the company's books at the end of the year. The extra money had shown up as a profit, which the company spent.

There is a method which can prevent such things from happening. Setup separate bank accounts for all items from which the contractor receives payment on a daily, or monthly, basis, and that the contractor makes payments on quarterly, or even annually. Some items might fit into this category are certain taxes, insurance of all kinds, equipment maintenance expenses, licenses, yearly fees, dues, etc. On a monthly basis, write checks for each of these items in the amounts you charge clients for those items. Deposit those checks in the separate savings account designated for each item. Monthly profit-and-loss financial statement will now include these costs because you wrote a check for them. Also, when the bill finally arrives, there will be the money in an account to pay it.

A third item about money discipline can almost go without saying, except it is too important not to mention. A contractor should bill every chance they get, and then be a regular "tiger" about collecting their receivables as quickly as possible.

Some contractors feel embarrassed to bill their job. They feel even more embarrassed to hound their clients for prompt payment. These contractors will not last very long in this business unless they take on an aggressive billing and collection policy. Very few days should go by without bills of some sort going out, and calls being made about receivables. A client might require a certain period of time to pay a bill. Send the bill right away and the time period begins sooner, which means the time when they will pay your company will come sooner, too.

A fourth item about money discipline involves a moneymaking principle. A contractor should take advantage of every discount of 2 percent or more, without fail, and 1 percent if they are able. One can borrow money to take advantage of a 2-percent discount on supplies, and still save more than the interest charged on the borrowed money. At 1 percent it becomes almost break-even. But, if a contractor has the money, getting the discount will pay them a better return than having the money in a savings account.

Finally, here is a principle by which one should live both in business and in personal money discipline: *Do not spend all your cash.* There are those who believe in buying everything for cash and, thus being debt free. If they are so well healed that they can do that, of course, that is a good position to take. However, since almost no one is in this position, a better position to take is one which maintains a balance between debt and cash in CDs, or in Money Market accounts, or as cash reserve against the payments of the debt.

This business is so volatile wherein all companies should keep a cash reserve for an emergency situation. It will carry them through a potential downturn in their workload. If they use all of their cash to pay for equipment or other purchases, they will not have payments on a monthly basis, it's true; but, if things take a turn for the worse, they will not have the cash to survive. At that point, in a bad situation, they will be at the mercy of a lending institution to loan them money in bad circumstances. If they finance a part of the equipment, and keep a cash reserve, they can make the payments, and keep going during a bad time, without depending on a lending institution.

Remember, a contractor's long-term success will not just depend on his ability to make money but, also, on his ability to handle the money well.

Cash Flow Management

Simply put, cash-flow management is the process of planning, budgeting, measuring, and controlling, the cash that flows into, and out of, your business.

The tools that follow are just that, tools. They cannot replace your sound business judgment, and you must always consider them in relation to one another. Often, one item viewed in and of itself can be quite misleading.

These tools do not replace the more formal business barometers, such as financial statements; current, quick, liquidity ratios, etc., which are familiar to your banker and accountant. Rather, we designed these tools for you to track and to display data which must be tracked in order for you to run your business successfully on a day-to-day basis. In this context, implement and use them in conjunction with the more formal business barometers, along with your overall business plan. Although I highly recommend using a spread sheet computer program to prepare the following ratios, graphs, and charts, it is not essential. They can be set up on an accounting pad and an 8 1/2" x 11" graph pad.

The following are some of the reasons for, and the goals of, a cash-flow management system:

- To provide a plan that reduces, or even eliminates, emotional and impulsive financial decision making (See section on page 241).

- To eliminate "surprises." There is no such thing as a "good" surprise in this business.

- To provide realistic goals. Goals should stretch you, yet be attainable.

- To measure your progress toward your goals.

- To motivate you and your people to attain your goals and your plans.

- To provide early warning of potential problems.

- To establish an historical—versus hysterical—track record for future planning.

- To identify past trends and ratios influencing your business, and then to plan accordingly.

- To accumulate consistent and accurate data.

- To measure your ability to plan, and to control your cash.

- To help you with future cash-flow management. (If you never start, how can you improve?)

The first tool is the ***Multi-Year Ratio and Trend Analysis*** (see illustration on page 254). The purpose of this tool is to identify historical trends and ratios, which you can use for current budgeting.

You begin to formulate the analysis by collecting the last three or four end-of-the-year financial statements and tabulating the income statement data. Do a separate sheet for Parts I-III and for Part VIII on the analysis form. Analyze all divisions of your company which comprise more than 20 percent of the company's gross sales, then if you deem it necessary, for divisions under the

20 percent. Different divisions — construction, maintenance, etc. — will have differing ratios and percentages.

Be aware that the consistency of categories of data is the key ingredient needed in order to obtain useful trends and ratios. It is better to have no multi-year ratios and trends available for your use, than it is to have ones which define overhead with all equipment costs one year and without those costs another year, for instance. Do not get hung up on all the small details when preparing this analysis. However, do write down notes as to how you arrived at your figures, and how you defined your categories.

The second tool is the *Yearly Cash Flow Budget Projection* (see illustration on page 255). This tool will help you set cash-flow goals for an upcoming period of time and it will also help you identify potential cash-flow shortfalls.

You begin by looking over our illustration to see if there are any categories you need to add. As you begin, you will need to remember this is your best educated guess for the next twelve months, and that you are beginning a process which you will refine and improve constantly as you gain experience over the months and years ahead. The important thing is to start the process and, then, to be consistent in monitoring and following it as much as possible. You will also want to be as realistic as possible with your projections—not too optimistic or pessimistic. You will also need to keep in mind uncontrolled growth is as dangerous as a raging river above flood stage. As I have mentioned before, but want to reiterate, do not, I repeat, *do not* grow beyond your ability to control the flow of the business and its cash.

I suggest you prepare a separate projection for each division that accounts for more than 20 percent of your company's gross sales or, if you deem it necessary, for divisions doing less than 20 percent. If you have division managers, allow them to prepare and to submit their own projections first. Combine all division budgets into a combined budget for the entire company once you all agree about individual division projections.

If possible, allocate overhead directly to divisions incurring those costs. For example: How much time does the office staff usually spend on each division? Companies with over two million in gross sales per year should seriously consider having their office staff submit timecards which indicate how much time went to support each division. You need to keep in mind that this budget includes all moneys paid out, but that it does not include depreciation.

All revenue sources and expenses should be included in the yearly cash-flow budget. The cash-flow excesses or deficits are excesses or deficits for each month while the cash-on-hand is an actual checkbook balance goal.

A third tool in this concept is what I call the *Periodic Recap Sheet* (see illustration on pages 256 and 257). It will help provide a periodic comparison of actual expenses to budget projections.

You will notice the categories are essentially the same as for the multi-year trend, the ratio analysis, and the yearly cash-flow budget projection. If necessary, you can amplify it to include—and to compare—all overhead categories.

You begin by preparing a recap for each division as well as a consolidated recap, as you deem necessary. Prepare recaps monthly by the second workday of each month, for the previous month. Line 5B should equal your checkbook balance. Receivables and payables can be included as in items 6 and 7 on the illustration. The subtotal below line 7A on the illustration should provide an indication as to the cash flow available to meet payroll during the next 30 days. Use receivables over 30 and 60 days past due, and the ratios in item 8, for analysis and charting, which we will cover later on in this section.

These percentages and ratios are meaningless unless you consider them in the context of the overall business picture. For instance, a material-to-labor ratio may be very low one month because you purchased the material one month, while you did not expend the labor to install it until the next month. Consequently, you must consider the yearly aggregate ratio with the monthly one.

I also want to show you some graphs and charts that I have found to be helpful. They can be very useful and powerful if used consistently. You can display vast amounts of crucial data, as well as vital trends, on a simple 8 1/2" x 11" sheet of graph paper. In the following paragraphs I will give you some suggested items to track.

The first graph is of **Receivables** (see illustration on page 258). Chart this graph, including all categories such as current, over 30 days, over 60 days, and a total category. Your goal will be to have the trend show all graphed lines "trending" toward the current category. If the trend is in the opposite direction, then you need to analyze why that is the case, and take corrective action.

The second graph is of **Payables** (see illustration on page 259). It is identical to the receivables graph.

The third graph is of the **Receivables to Payables Ratio** (see illustration on page 260). The graph will help you monitor the flow of cash into and out of the business, and motivate you to keep your billing current.

Chart this graph weekly, as well. You divide the current receivables by the total payables, excluding payroll. Each dollar of payables should have at least one dollar of current receivables to cover it. Establish a goal line. I recommend a goal of at least 1.5. This means you have $1.50 in current receivables for every $1 of payables. After establishing such a goal, strive to keep above the goal line. If you dip below it, submit additional billings or get past due receivables paid to you. If you experience cash-flow shortages while keeping above the 1.5 line, adjust the goal line upwards.

The fourth graph is of the *Material to Labor Ratio* (see illustration on page 261). This graph will help you monitor field performance, and compare the reality of what is happening in the field to your budget projections.

Chart this graph on a monthly and year-to-date basis. Divide the amounts of actual cash paid for material (cost of sales) by the cost of direct labor during the same period. Then compare the ratios to the budget projections, and make appropriate budget adjustments, if the two are consistently out of line. You will also want to keep this ratio in mind when you are preparing future bids.

The fifth graph is of *Direct Labor Budget In Comparison To Actual Labor* (see illustration on page 262). This graph is very important because it also ties into your estimating strategy. It will tell you if you are making a budget for labor which will recover your overhead, or if you will need to make adjustments to your strategy.

Chart this graph weekly. If you find great variance, make adjustments to your overhead-recovery percentages in your bids, or else you must reduce your overhead.

The final graph is of *Projected Gross Sales In Relationship To Completed Work* (see illustration on page 263). This graph will represent actual billings in relation to the sales goals which you have made.

This graph should be charted weekly or monthly, depending on when you do your billings. Although it is not as important to your estimating strategy as the previous graph, it can be a visual motivator, as well as a reference for future projections.

After reading about all these projections, budgets, and graphs, you may already be dismissing them as more trouble than they are worth. If this is the case, I want you to reconsider. Not every one of you will need to do all of these. But, each of you needs to do some of them. Once you begin to see their effect on your ability to manage your cash and your company, I believe you'll want to use more and more of them.

Use these graphs as they are, or adapt them and make graphs you like better. If you use them, any of them, you will have succeeded already in running your company more efficiently.

	FY 88 BUDGET	%	FY 87	%	FY 86	%	FY 85	%	FY 84	%	$ AVE.	% AVE.
											4	
I. GROSS SALES	1,120,000	100.0%	968,676	100.0%	897,491	100.0%	779,149	100.0%	762,229	100.0%	851,886	100.0%
II. COST OF SALES												
MATERIAL	209,000	18.7%	163,538	16.9%	158,992	17.7%	136,579	17.5%	160,203	21.0%	154,828	18.2%
LABOR	175,000	15.6%	149,928	15.5%	130,666	14.6%	124,995	16.0%	128,072	16.8%	133,415	15.7%
BURDEN 31.0%	54,250	4.8%	23,975	2.5%	22,580	2.5%	19,135	2.5%	15,472	2.0%	20,291	2.4%
EQUIPMENT	26,000	2.3%	24,333	2.5%	26,844	3.0%	23,372	3.0%	20,664	2.7%	23,803	2.8%
EQUIP. RENTAL	4,000	0.4%	2,989	0.3%	4,390	0.5%	2,481	0.3%	6,003	0.8%	3,966	0.5%
SUB-CONTRACTORS	302,400	27.0%	276,695	28.6%	236,721	26.4%	221,717	28.5%	130,746	17.2%	216,470	25.4%
MISC.	7,800	0.7%	7,197	0.7%	6,099	0.7%	5,475	0.7%	4,652	0.6%	5,856	0.7%
TOTAL	778,450	69.5%	648,655	67.0%	586,292	65.3%	533,754	68.5%	465,812	61.1%	558,628	65.6%
III. GPM	341,550	30.5%	320,021	33.0%	311,199	34.7%	245,395	31.5%	296,417	38.9%	293,258	34.4%
IV. OVHD. TOT.	262,840	23.5%	295,670	30.5%	258,090	28.8%	234,484	30.1%	274,685	36.0%	265,732	31.2%
V. OPERAT. INC.	78,710	7.0%	24,351	2.5%	53,109	5.9%	10,911	1.4%	21,732	2.9%	27,526	3.2%
VI. OTHER INC.	6,000	0.5%	6,579	0.7%	3,063	0.3%	6,074	0.8%	2,924	0.4%	4,660	0.5%
VII. PRE-TAX EARN.	84,710	7.6%	30,930	3.2%	56,172	6.3%	16,985	2.2%	24,656	3.2%	32,186	3.8%
VIII. RATIOS												
MAT/LAB	1.19		1.09		1.22		1.09		1.25			
COS/GS	69.5%		67.0%		65.3%		68.5%		61.1%			
OVHD/GS	23.5%		30.5%		28.8%		30.1%		36.0%			
NET/GS	7.0%		2.5%		5.9%		1.4%		2.9%			
GP/GS	30.5%		33.0%		34.7%		31.5%		38.9%			
GS GROWTH	15.6%		7.9%		15.2%		2.2%		N/A			

Multi-Year Ratio/Trend Analysis

YEARLY CASH FLOW BUDGET

	JAN	FEB	MAR	APR	MAY	JUN	JUL	AUG	SEP	OCT	NOV	DEC	TOTAL
INCOME:													
CASH ON HAND	20,000	41,445	62,889	84,334	105,778	127,223	148,667	170,112	191,556	213,001	234,445	255,890	
COLLECTIONS	200,000	200,000	200,000	200,000	200,000	200,000	200,000	200,000	200,000	200,000	200,000	200,000	2,400,000
COST OF SALES:													
MATERIAL	76,000	76,000	76,000	76,000	76,000	76,000	76,000	76,000	76,000	76,000	76,000	76,000	912,000
LABOR	38,000	38,000	38,000	38,000	38,000	38,000	38,000	38,000	38,000	38,000	38,000	38,000	456,000
LABOR BURDEN	12,540	12,540	12,540	12,540	12,540	12,540	12,540	12,540	12,540	12,540	12,540	12,540	150,480
SUB-CONTRACTORS	10,000	10,000	10,000	10,000	10,000	10,000	10,000	10,000	10,000	10,000	10,000	10,000	120,000
EQUIPMENT	5,000	5,000	5,000	5,000	5,000	5,000	5,000	5,000	5,000	5,000	5,000	5,000	60,000
EQUIPMENT, RENTAL	6,000	6,000	6,000	6,000	6,000	6,000	6,000	6,000	6,000	6,000	6,000	6,000	72,000
TOTAL COST OF SALES	147,540	147,540	147,540	147,540	147,540	147,540	147,540	147,540	147,540	147,540	147,540	147,540	1,770,480
GROSS PROFIT	52,460	52,460	52,460	52,460	52,460	52,460	52,460	52,460	52,460	52,460	52,460	52,460	629,520
(OVERHEAD)													
ADVERTISING	500	500	500	500	500	500	500	500	500	500	500	500	6,000
DONATIONS	8	8	8	8	8	8	8	8	8	8	8	8	100
DUES/SUBSCRIPTIONS	83	83	83	83	83	83	83	83	83	83	83	83	1,000
INSURANCE:OFF EQ,LIFE(KEYMAN)	1,000	1,000	1,000	1,000	1,000	1,000	1,000	1,000	1,000	1,000	1,000	1,000	12,000
INTEREST & BANK CHARGES	2,702	2,702	2,702	2,702	2,702	2,702	2,702	2,702	2,702	2,702	2,702	2,702	32,426
DOWNTIME	667	667	667	667	667	667	667	667	667	667	667	667	8,000
LABOR BURDEN (DOWNTIME) 34.0%	227	227	227	227	227	227	227	227	227	227	227	227	2,720
OFFICE SUPPLIES	208	208	208	208	208	208	208	208	208	208	208	208	2,500
PROFESSIONAL FEES: CPA	125	125	125	125	125	125	125	125	125	175	125	125	1,500
PROFESSIONAL FEES: LEGAL	75	75	75	75	75	75	75	75	75	75	75	75	900
PROFESSIONAL FEES: MISC	575	575	575	575	575	575	575	575	575	575	575	575	6,900
MORTGAGE/RENT	1,350	1,350	1,350	1,350	1,350	1,350	1,350	1,350	1,350	1,350	1,350	1,350	16,200
SALARIES - OFFICE	11,667	11,667	11,667	11,667	11,667	11,667	11,667	11,667	11,667	11,667	11,667	11,667	140,000
SALARIES - OFFICER	4,500	4,500	4,500	4,500	4,500	4,500	4,500	4,500	4,500	4,500	4,500	4,500	54,000
LABOR BURDEN (OFFICE) 14.0%	1,633	1,633	1,633	1,633	1,633	1,633	1,633	1,633	1,633	1,633	1,633	1,633	19,600
SMALL TOOLS & SUPPLIES	250	250	250	250	250	250	250	250	250	250	250	250	3,000
TAXES - PROPERTY & BUSINESS	150	150	150	150	150	150	150	150	150	150	150	150	1,800
TELEPHONE	800	800	800	800	800	800	800	800	800	800	800	800	9,600
TRAVEL & ENTERTAINMENT	250	250	250	250	250	250	250	250	250	250	250	250	3,000
UTILITIES	200	200	200	200	200	200	200	200	200	200	200	200	2,400
YARD EXPENSE	225	225	225	225	225	225	225	225	225	225	225	225	2,700
OVHD. VEHICLE PAYMENTS	475	475	475	475	475	475	475	475	475	475	475	475	5,700
OVHD. VEHICLE GAS/MAINT/ALL ELSE	525	525	525	525	525	525	525	525	525	525	525	525	6,300
RADIO SYSTEM	350	350	350	350	350	350	350	350	350	350	350	350	4,200
LICENSES, BONDS	50	50	50	50	50	50	50	50	50	50	50	50	600
EDUCATION	125	125	125	125	125	125	125	125	125	125	125	125	1,500
UNIFORMS & HARD HATS	45	45	45	45	45	45	45	45	45	45	45	45	540
COMPUTER	250	250	250	250	250	250	250	250	250	250	250	250	3,000
BAD DEBTS	2,000	2,000	2,000	2,000	2,000	2,000	2,000	2,000	2,000	2,000	2,000	2,000	24,000
MISCELLANEOUS	50	50	50	50	50	50	50	50	50	50	50	50	600
TOTAL OVERHEAD	31,015	31,015	31,015	31,015	31,015	31,015	31,015	31,015	31,015	31,015	31,015	31,015	372,186
CASH FLOW EXCESS/(DEFICIT)	21,445	21,445	21,445	21,445	21,445	21,445	21,445	21,445	21,445	21,445	21,445	21,445	257,334

Yearly Cash Flow Budget Projection

	JANUARY THRU MARCH				Y-T-D	Y-T-D	Y-T-D	Y-T-D
	BUDGET	ACTUAL	VARIANCE	%	BUDGET	ACTUAL	VARIANCE	%
1. SALES	200,000	193,000	(7,000)	100.0	200,000	193,000	(7,000)	100.0
2. COST OF SALES:								
MATERIAL	66,500	67,832	(1,332)	35.1	66,500	67,832	(1,332)	35.1
LABOR	35,000	31,020	3,980	16.1	35,000	31,020	3,980	16.1
LABOR BURDEN	10,850	8,966	1,884	4.6	10,850	8,966	1,884	4.6
EQUIPMENT	12,500	13,582	(1,082)	7.0	12,500	13,582	(1,082)	7.0
EQUIPMENT, RENT.	1,000	490	510	0.3	1,000	490	510	0.3
SUBS-CONTRACTORS	2,000	2,200	(200)	1.1	2,000	2,200	(200)	1.1
MISC.	0	0	0	0.0	0	0	0	0.0
TOTAL	127,850	124,090	3,760	64.3	127,850	124,090	3,760	64.3
3. GROSS PROFIT	72,150	68,910	3,240	35.7	72,150	68,910	3,240	35.7
4. OVERHEAD TOTAL	48,000	46,639	1,361	24.2	48,000	46,639	1,361	24.2
A. SALARY OFFICE	14,750	13,925	825	7.2	14,750	13,925	825	7.2
B. SALARY OFFICER	6,000	5,500	500	2.8	9,000	5,500	3,500	2.8
5. OPERATING INCOME	24,150	22,271	(1,879)	11.5	24,150	22,271	(1,879)	11.5
A. OTHER INCOME (EXP)		0				0		
B. TOT. OP. INCOME/(LOSS)		22,271				22,271		
6. TOTAL PAYABLES		128,884		66.8				
SUB-TOTAL		(106,613)						
7. RECEIVABLES:								
A. CURRENT		187,300		97.0				
SUB-TOTAL		80,687						
B. OVER 30		33,457		17.3				
SUB-TOTAL		114,144						
C. OVER 60		2,500		1.3				
SUB-TOTAL		116,644						
8. RATIOS:								

Periodic Recap Sheet - Part 1

A.	MATERIAL/LABOR	1.90	2.19	1.90	2.
B.	CUR REC/TOT PAY		1.45		
C.	CUR REC/TOT REC		83.9%		
D.	OVER 30/TOT REC		15.0%		
E.	OVER 60/TOT REC		1.1%		

Periodic Recap Sheet - Part 2

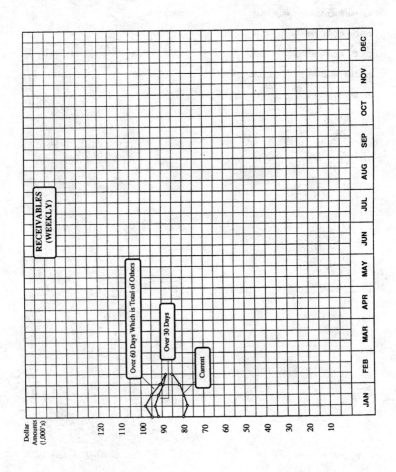

Receivables - Weekly

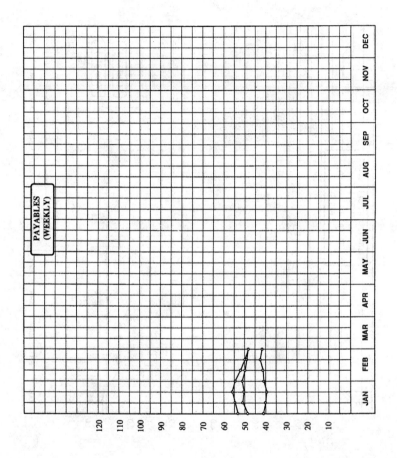

Payables - Weekly

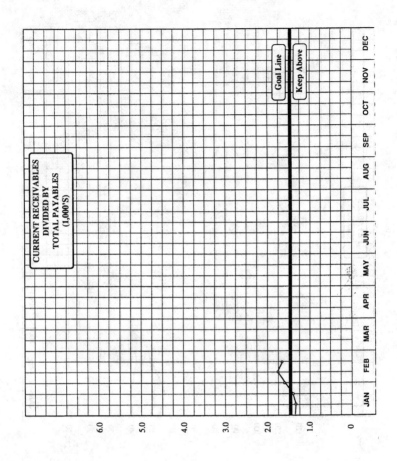

Receivables to Payables Ratio (Current Receivables Divided by Total Payables)

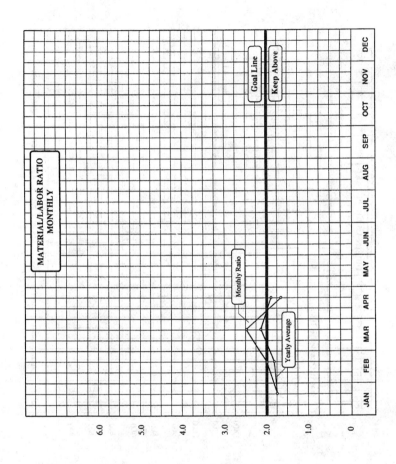

Material to Labor Ratio - Monthly (Materials Cost Divided by Labor Cost)

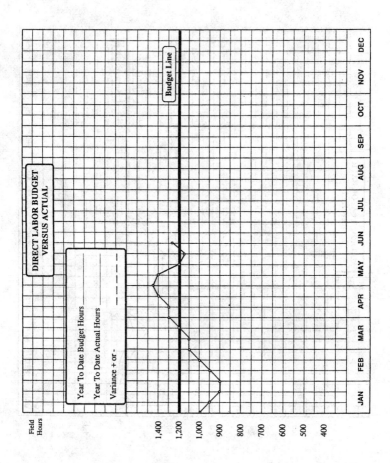

Direct Labor Budget In Comparison To Actual Labor

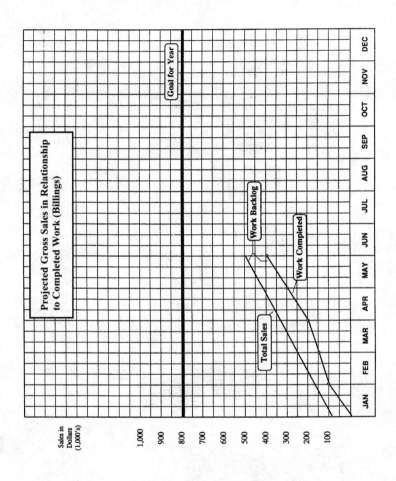

Projected Gross Sales In Relationship To Completed Work (Billings)

Corporations, Families, Partnerships, and Other Horrors

Over the sixteen years I have been consulting with contractors I have seen several misunderstandings and abuses of the methods by which a company is run and organized. The most common problems have to do with the way people operate corporations.

When corporations are made up of members of one family, it is a problem. Usually these family corporations ignore the concepts and benefits a corporation brings, and run their business as they would run their family. The dad or mom, or whoever started the business, runs it all and makes all the decisions. They do not appoint a Board of Directors, and hold no meetings. Stockholders never vote. To make things worse, all the dynamics of the family and its history play out in the business. If Dad has been an intimidating father, it continues to be that way in business. If brothers or sisters have felt one or the other has received favoritism in the home, the same is seen and felt in the business. The owner of the business may give stock to different family members at different times when the stock has a different value. One family member comes into the company when the worth of the company is down. They work hard and bring up the worth of the company through profits. Several years later, Dad gives another family member an equal amount of stock with a greater value due to the hard work of the first family member.

When I get involved in such situations I recommend the following:

1. Establish the worth of the company on a regular basis. Family members should buy any stock at its current worth. The terms (loan, interest, pay increases to be able to pay for it) should be as generous as possible. This concept has several benefits. One, family

members will respect the ownership if they are required to pay for stock, rather than having it given to them. Second, if Mom and Dad sell the stock, the revenue flows into their future net worth (or retirement) so what they pass on to their kids will be from the net worth, not the worth of the company. If Dad gives stock only to the kids who are interested in the company, then those who aren't interested feel cheated. If the stock is passed out evenly to all the kids, then those who work the company either make or break the inheritance of the kids who are not interested or who do not work in the company. Both of these scenarios can cause great problems and family stresses over the stock of the company. Third, those who buy in pay the increased price based on the current price of the stock when they buy in, thus eliminating any concern from those who might have come in earlier.

2. Appoint a Board of Directors who can advise and mediate family struggles over the business decisions. Appoint unbiased, multi-aged people whom all the members of the family respect. This board should meet three to four times a year to help the corporation make decisions and resolve any conflicts.

3. Dad or mom should do two things as the major stockholder or president of the company. First, acknowledge that the perpetuity of the company will occur only if they are willing to give up the reins. Second, establish a plan with the other family members as to how and when the control will shift. My greatest admiration has been for those people who have known how to bow out of a company with grace and honor.

Another problem occurs when people feel that being a stockholder is what determines a persons salary. Here is what happens. Two or three people are the major stock holders. Often times they own equal or nearly equal amounts of stock. Because they are nearly equal owners, they feel they should all have the same salary. This really becomes a travesty of corporate justice when one has the responsibility of running the company, another runs the field, and another may be a foreman on the job. (That may sound like a wild example, but I have seen it.) It is just as bad when people think that stock ownership entitles them to a certain kind of job. These people may or may not be qualified for the job, but because of their ownership, they insist on the position.

I remind people in this situation of the following concepts:

1. Stock ownership entitles you to a share of the profit based on your percentage of stock. It does not entitle you to a certain job at a certain salary. Just because I own stock in United Airlines does not mean I should be flying one of their jets.

2. People should be paid different salaries based on the responsibility they have and the industry standard for the job they perform. This means when there are multiple stockholders, there must be a difference in salary levels, because they cannot all be performing the same job.

Another problem: Let's say the owner of a company who has struggled along for years has finally hired a good person to work for them. The owner feels that in order to hire and keep this person, they need to give them a part of the company.

I recommend you do the following:

1. Do not give the stock away. Sell it to them based on the methods and for the reasons previously discussed.

2. Go over the understandings so there will be no misunderstanding. I have also just mentioned concerns regarding salaries and positions and stock ownership.

3. Set up a waiting period between discussing and agreeing upon the sale of stock, and the actual sale and transfer of stock. I like to equate the selling of stock to the dating process. First there is a period of time for social dates which can be equated to them working for you. Then there is the engagement period when a commitment is made but the relationship has an opportunity to be evaluated under the strains of the commitment. This is the waiting period of time I just recommended. Then comes the marriage with the actual exchange of stock.

Yet another problem: Partnerships. Partnerships are just like marriages. The same relationship dynamics and work which a marriage requires will be required in a partnership. Before you enter into a partnership you should go through the three stages I just described — dating, engagement and then marriage. Also, be sure you are ready to commit to the work which goes into a partnership.

One last bit of advice: No matter which way you sell the stock or enter partnerships, make sure you have in writing a good method to either buy back the stock or dissolve the partnership. It may be unromantic to have a prenuptial agreement, but this *is* business.

The Risks of Growing Your Business

Whenever I stand in front of a room full of contractors, I always congratulate them for picking one of the most difficult businesses to run and manage in the world. Every variable which can sink a business exists in the contracting business. There are variable site conditions, variables in weather, variable owners and architects, a variable workforce and variable amounts of material to labor to equipment usage on each job, from which you must recover a variable overhead.

If you had gone into the manufacturing business you could figure to the penny what it would cost to manufacture a certain product. To this, you could add overhead and profit and come up with a sales price. Once this was established, you could go out and make a zillion of the product and each would cost and sell for the same. But in contracting, you are constantly making a different product depending on the job, the site conditions, and what the owner or architect want on the job. That's what makes this business tough, and the reality will never, ever, be changed.

Statistics say that two out of three new contracting business never make it to the third year. Less than twenty percent make it to the tenth year. The only business that is more risky is the restaurant business.

I say all of this to set the stage for discussing the risks of expanding your business. The first thing I want to discuss is what I call the farmers and the ranchers. Farmers are like Ma and Pa contracting companies. Pa goes out and bids jobs, gets the contractors, organizes the jobs and acts as operation manager and supervises the crews. Ma (or a secretary/bookkeeper) stays in the office and answers the phone, bills jobs and handles the money. Well over 80 percent of the contracting in this country is done by farmers.

Ranchers are contractors who hire other farmers to be a part of their ranch. Ranchers spend almost all their time supervising the farmers who work for them and do the bidding, supervising and office work. They no longer have time to do this kind of work themselves because they are busy motivating and supervising the farmers.

I also believe all of us are born to either be primarily a farmer or a rancher. Most of the contractors in the U.S. are farmers, but there are a few gifted and able to be ranchers. The problem is some of us have not considered or accepted who we are or are not.

I rarely get to hear other speakers at a convention because I am already on an airplane to get to my next engagement. But every once in a while I do get to go to a banquet or some other function. I have seen the organization bring in one of its most successful members or one from another part of the country. He or she is usually a big rancher running a big company, and brings in slides of all their big jobs, their beautiful office, shop layout and all their programs they have in place. When I see this I say "God bless this speaker, for they mean well, but they are bankrupting some people in this room."

Why do I say this? Because I look out on a room where 80 percent of the people are farmers, most of whom should stay this way because they were born this way. But now a lot of them have stars in their eyes as they see what this big rancher is doing in their business. I can hear them thinking they want to be just like that rancher someday. However, they do not have what it takes to become a rancher and when they attempt to become one they will lose a lot of money if not their company.

You see, when you started your company you were put into a hopper with two other contractors. Over three years of the hopper going round and round, the other two quit or were thrown out of the hopper. You got to stay. Your risk of failure as a farmer once you have passed that first test is quite small. However, if you want to grow significantly and become a rancher, you are put back into a hopper with two other farmers who want to become ranchers. Once again the hopper will go round and round and you will be at risk of losing a lot of money and quitting your attempt, or being thrown out of the hopper and losing your business.

Before you ever decide to grow significantly you need to determine whether you are a farmer or a rancher.

The other problem with growth in this business is that it does not grow very evenly. Let me explain what I mean. A Ma and Pa business has a maximum amount of gross sales they can accomplish. When they hit the maximum they are doing the most they can with their overhead people and are making the maximum profit which can be made. To go to another level requires the addition of another overhead person. However, to add another overhead person to a company which only has two people on overhead requires the company to add fifty percent to their present sales to justify the additional overhead, and this growth must occur almost immediately. Three dynamics all begin to kick in at once. First, you have a new person to whom you must adapt, as well as the rest of the company. This person is often an unknown commodity as to their real talents and abilities. Second, you must expand your field capacity in people and equipment to handle this extra work almost immediately. Third, the additional payroll and equipment payments will put an added strain to your cash flow.

These dynamics can be approached from two different methods. The first is for you to work harder and longer hours to add some of the additional sales before the additional person is added. This will cut down on the risk and the losses if the expansion does not succeed. The second approach is to add the person and just accept the fact for at least half of the year you will take your profit and invest it into the additional overhead until sales kickup to the needed amount and this is simply an investment in your future growth.

What is Your Company Worth?

Probably not as much as you either think or want it to be worth. In fact, in most cases, contracting businesses are near the bottom of the list of the kinds of good businesses to build and sell. Here are just some of the reasons why this is so.

1. A lack of a loyal client base. Most contractors work either on a one-time basis with people or on a price-sensitive basis. Consequently, most contractors gain business by having a competitive price, not because an owner will use a company and regardless of price.

2. Contracting companies are usually personal businesses that revolve around the personality of you, the owner. What repeat business you enjoy primarily exists because of your relationship with a client. If I buy your company, that relationship may not continue because my personality does not mesh with the clients. Also, many of your employees fit your personality and work well with you because of your relationship. If I buy your company, my personality may not work well with your staff and I may lose a majority of them during the transition.

3. Contracting is so easy to get into from scratch; thus, there is usually not a strong motivation to buy an existing company. For a few thousand dollars someone can get cards and brochures printed and spend time getting on bid lists, etc.

4. Contracting is the second riskiest business to be involved in. You don't see the investors of Wall Street

rushing in to buy interest into any construction company. Most contractors I have met are in it for the love of the business, not because they are enjoying tremendous returns in their investments.

5. Contracting rarely has a "niche" in the marketplace. If you had invented a new widget and had a patent on it and created a market for widgets, your company would be worth a lot of money. But most contractors have no patents or widgets but rather are doing fairly typical things that almost any contractor could accomplish.

So, with all this in mind, what is your company worth? There are three areas of worth to be considered in placing a value on your company:

1. Assets. Assets come in many different forms: Cash in the Bank, Accounts Receivable, Prepaid Expense, Equipment, Supplies, Inventory, Furniture and Fixtures, Autos and Trucks, and Accounts Receivable. Many of these have value only after taking age, wear and tear, and general condition into account. Sometimes, you would not even be able to get the book value of equipment and inventories in a business sale. Therefore, it is very important to be brutally realistic in valuing assets to determine what your company is worth.

2. The company name and reputation or what is often referred to as "blue sky." This is what many owners who are selling think is worth a lot, but in reality is worth very little. First, sometimes there are skeletons in the closet and an ongoing name can be as debilitating as it is advantageous. Second, when the company changes hands, the loyalty and perceived reputation may not stay with the company that is under

new management. Every company is worth some blue sky if it is reputable and has been in business for at least five years. However, it will vary greatly depending on the type of business it is (design/build/hard bid/residential/commercial/maintenance/retail) and the type of area (city/rural/small town) in which it is located. I've seen as little as a $1,000 paid for blue sky, and as much as $100,000 (major company) paid for it.

3. Existing contracts. For a construction company these are worth very little. Who knows for sure if the expected profit will be there at the end of the job? Who knows whether that profit will exist because of a good bid or because you took it over and made it make that profit? Buying contracts is usually buying someone's risks. However, I feel construction contracts are worth the cost of obtaining them. For larger contracts, that cost should be around 1 to 2 percent of the contract. For smaller jobs, it could run between 3 to 5 percent.

Maintenance contracts, because of their ongoing, year-to-year renewal potential are worth more money. What I have been recommending is half the profit (after overhead) for three years. However, this is not paid out at the time of purchase. A list of contracts and payouts for each contract is made. As long as the client pays and remains a client, the payment is made. If at the end of a year or two, a client drops off the list and does not renew, the payout also drops off the list.

The other major concern is what will happen to you after the company is sold? If an outside investor or someone new to the industry buys it, they will want you to stick around for some period of time. If you sell to this kind of buyer, make up your mind to do it for your employee's and client's and buyer's sakes. If you don't want to stick around, sell to another contractor in your area who will probably pay less money but look at you as a way to grow his company. Most times, they will not want you around after the business is sold.

Another good source of buyers for your company may be your employees. They know the company, its systems, and clients and will want to see the company go on for the sake of their job.

Two Formats for a Financial Statement

The most important characteristic of a financial statement is the need to follow the same format used to generate the budgets. If it does not, it will be impossible to compare actual costs to budgeted costs to determine accuracy of the budget. The concern for how equipment costs and labor burden costs relate to overhead cause a need for separate accounting under the Indirect Cost of Sales account(s).

If the equipment costs and labor costs are all put in overhead, they will not line up with a contractor's overhead in the budget since most of these costs are arrived at through the estimate as job costs.

The "Simplified" format (page 281) is easiest and should be used by contractors who are either small or do not have sophisticated enough accounting systems to the "More Complicated" format. Here Indirect Cost of Sales is a single category.

The "More Complicated" format (page 282) divides "Indirect Cost of Sales" into two parts. "Indirect Costs - Equipment" and "Indirect Costs - Labor Burden." The equipment costs include such things as depreciation, interest on equipment loans, repairs, insurance on equipment, licenses for equipment and fuel. The Labor Burden Costs include such items as the company share of FICA, workman's compensation, unemployment insurance, liability insurance, vacation pay, holiday pay, and health insurance benefits for field people.

If indirect costs are divided this way, the contractor can make direct comparisons, after a period of time, between actual and budgeted costs for both equipment and labor burden to ensure cost factors are correct for the estimating process.

The second format is more complicated, but is the best format and a contractor should strive toward implementing this format. It is shown on page 282.

The first thing which needs to be pointed out is that both the equipment used in the field and the labor burden which applies to field labor is shown under costs of sales. This is done by what is called a journal entry which is inserted by the accountant. These figures are taken or applied from two different sources.

The labor burden figure is applied by using the same percentage which is used in estimating. If this percentage, for example, is 30 percent then 30 percent of the field labor for the month is inserted in the cost of sales as the labor burden cost for this particular month.

The equipment figure is arrived at by keeping track of the use (rental) of the contractor's equipment on the jobs for a given month. This is done by the field people who keep track of the use of equipment on the time cards (see the time card example on page 157). These hours or days of use are then posted to some spread sheet program on a regular basis. At the end of a month, a total usage of each type of equipment can be totaled. The total hours or days of use are then multiplied times the rental cost which the contractor uses in estimating (see page 443). This figure is the amount of the journal entry for the costs of the use of the contractors equipment which is owned by the company.

The next important concept to note is the two areas of indirect cost of sales. There is one for labor burden and one for equipment.

Under the indirect cost of sales for labor burden, the first thing which must be done is the crediting of labor burden where it is journal entried elsewhere. The labor burden which was put in cost of sales is a journal entry credit as well as the labor burden put against the mechanic (if a contractor has one), and the labor burden put against overhead salaries (see overhead on this example) which has been put in at the percentage arrived at during budgeting. These three credits have then removed all of the journal entries put in the different categories at the budgeted percentage. Following those credits, the accountant keeps track of the exact amounts of the labor burden costs. Those debits should offset the credits for a zero or near zero balance of this section of indirect costs. If this is the case, then the percentages being used in estimating are accurate. If there is a plus or minus as a total in this category, the labor burden percentages need to be adjusted.

A negative total in the Indirect Costs of Sales/Labor Burden section means the contractor's labor burden is actually less than the percentages being used in the estimates. If it is a positive total, the contractor is not covering labor burden costs with the percentages being used, and needs to raise those percentages accordingly.

Again, the equipment put in cost of sale for the contractor's use of equipment is credited in this area. Also, the vehicles which are used by overhead people are rented to overhead on a monthly basis and inserted into overhead as a journal entry and credited here. The actual cost of maintaining the equipment, including the cost of a mechanic and a shop (if the contractor has one), is accounted for at exact cost.

The Indirect Cost of Sales/Equipment section becomes a profit and loss on the contractor's equipment usage (much as a separate rental company). If there is a negative figure at the bottom of this section, the contractor is making money on his equipment. This may be due to excellent care of the equipment and in keeping the repairs of the equipment down. It can also mean due to the excellent care of equipment, they are getting more extended usage out of the equipment than was actually calculated in costing the equipment (see page 443).

Otherwise, if this sections total shows a plus figure, it means there are more costs toward the equipment than are being recovered through usage. This means the contractor needs to consider charging more for the use of the equipment by raising the prices he is using on his estimates. Or, he may be under-utilizing the equipment and, since its costs continue, it is not generating the necessary revenue. This means the contractor should sell some of the equipment or find ways to use it more often. Or, the equipment is being abused by the field personnel, which raises the cost of repairs, and the contractor needs to investigate how the field is treating the equipment.

The overhead section is identical to the one which is used for budgeting purposes. Any downtime and its labor burden is tracked, labor burden for overhead salaries is costed in the same matter as was budgeted, as is the overhead vehicles. This gives a contractor the opportunity to keep his overhead within the amount budgeted, and to see where any variances from budget might occur.

This second format of a financial statement is preferred because it gives the contractor every opportunity to control his budgets and expenses in EVERY category.

SIMPLIFIED SAMPLE
PROFIT AND LOSS STATEMENT FORMAT

Sales $_____

Cost of Sales:
Material and tax
Labor
Subcontractors
Rental equipment

Total Cost of Sales: $_____

Indirect Cost of Sales:
Payroll taxes
Holiday and vacation pay
Employee health insurance
Liability insurance
Equipment depreciation
Equipment maintenance
Gas and oil

Total Indirect Cost of Sales $_____

Overhead
(Same as on the other illustration)

MORE COMPLICATED
SAMPLE PROFIT LOSS STATEMENT FORMAT

SALES: $_____

DIRECT COST OF SALES:
    Material
    Labor
    Labor Burden (at bid percent - Journal Entry)
    Rental Equipment
    Company Equipment (from time cards at bid rates - Journal Entry)
    Subcontractors

TOTAL DIRECT COST OF SALES: $_____

GROSS PROFIT $_____

INDIRECT COST OF SALES/EQUIPMENT
    Company Equipment (Journal Entry Credit from Cost of Sales)
    Company Equipment/Overhead (Journal Entry Credit from Overhead Recovery)
    Depreciation
    Interest (on Equipment Loans)
    Insurance (on Equipment)
    Maintenance and Repair
    Gas and oil
    Licenses
    Mechanic (if have one)
    Mechanic Labor Burden (if have one)
    Shop Rent (if have one)
    Shop Utilities (if have one)
TOTAL INDIRECT COST OF SALES/EQUIPMENT
(Profit/Loss on Equipment) $_____

INDIRECT COST OF SALES/LABOR BURDEN:
    Labor Burden (Journal Entry Credit from Cost of Sales)
    Labor Burden (Journal Entry Credit from Mechanic)
    Labor Burden (Journal entry Credit from overhead)
        F.I.C.A.
        F.U.T.A.

S.U.T.A.
Workmean's Compensation
Liability Insurance
Vacation Pay
Holiday Pay
Field Health Insurance
Other

**TOTAL INDIRECT COST OF SALES/LABOR BURDEN
(Profit/Loss on LB)** $_____

OVERHEAD:
Advertising
Depreciation (Office Equipment and Furniture)
Donations
Dues and Subscriptions
Insurance (Office Items and Health/Life)
Interest and Bank Charges
Downtime
Labor Burden (Downtime)
office Supplies
Professional Fees
Rent
Salaries - Office
Salaries - Officer
Labor Burden (Office)
Small Tools and Supplies
Taxes - Business
Telephone
Travel and Entertainment
Utilities
Yard Expense
Overhead Vehicles
Radio Systems
Miscellaneous
Licenses, Bonds
Education
Uniforms and Hard Hats
Computer
Bad Debts

TOTAL OVERHEAD $_____

NET PROFIT/(LOSS) $

*More complicated Sample Profit and Loss Statement
Format, Page 2*

Understanding and Using a Financial Statement

Financial statements are one of the most important tools a contractor can use in managing their business. However, it remains one of the most neglected tools simply because many do not know how to use them or what they are saying about their company. This section will attempt to point out the many uses of the financial statement.

The first part of the financial statement is the balance sheet. It is the part of the financial statement which reconciles the worth of a company by balancing the liabilities (money owed) with the assets (money in the company or owed to the company). It can best be evaluated by looking at some general indicators which this sheet will tell a contractor.

The first one is the accounts receivable in relationship to the accounts payable. If the accounts payable is greater than the accounts receivable, then the company is what is called "upside down" and unless they have cash to make up the difference, are in serious trouble. A company with that problem is on its way to bankruptcy unless it can make enough profit within twelve months to reverse this condition. A company that operates much beyond a year in this condition will almost always go bankrupt.

The ideal relationship of accounts receivable to accounts payable is a 1.5 to 1 ratio. For every $1.50 that is owed the contractor, they owe a $1.00. This kind of ratio shows that the contractor is doing a good a job of collecting receivables and paying its payables in a timely manner. If the amount owed a contractor is greater than the above-mentioned ratio, with more money owed the contractor than they owe, it is a sign that the receivables are not being collected in a timely manner or that the payables are being paid too rapidly. In order to maintain a good cash flow, a contractor must walk a fine line with his suppliers so that he is not paying them too quickly in relationship to how quickly he is being paid. The accounts receivable to accounts payable ratio can help them maintain that proper balance.

Another indicator of the health of a contractor shown on the balance sheet is the amount of short term loans which are due. Under the liabilities section of the balance sheet there is a section which shows the total loans a contractor has taken out. A portion of those loans are due within twelve months, so they are called "short-term loans due." If this amount is larger than one-fourth of the total amount of the loans, the contractor should look into refinancing those loans into longer term loans. The exception to this concept occurs when a contractor has a loan from an unsecured line of credit. That loan should not exceed either six weeks of payroll or the amount of difference between the receivables and the payables.

The last indicator to examine is the amount of retained earnings a contractor has in the company. Because the construction business is so risky, it is recommended that a contractor not keep a lot of his money in the company. The retained earnings are an indication of how much money a contractor is using of his own to finance equipment, receivables which are not being paid and for the weekly payroll. If the retained earnings are too high, it is an indication that a contractor has become his own bank. While this may seem an ideal position to be in and is certainly admirable, it is also risky. It is better to use the company as a means of generating cash to invest in other and more diverse investments and let the bank do its job in loaning money for equipment and a line of credit to do business.

One of the reasons for this concept is that a contractor can develop and maintain an ongoing and long term relationship with his bank. The worst time for a contractor to go to a bank is when he is in financial trouble and needs money. Without a track record and experience with a bank, a bank will be very hesitant to loan money, especially if the contractor is in financial trouble. However, when contractors use a bank to finance their equipment and payroll, even when they do not need to because they could use their own retained earnings, they develop and maintain that relationship. This allows them to have their retained earnings as a backup in case there is trouble, rather than having to go to a bank when they are in trouble and being denied the necessary help.

Another important indicator on a financial statement is when a contractor has several divisions or profit centers. These should be kept as separate profit and loss reports with a consolidated statement of all these divisions into an overall company Financial Statement.

During good economic times in which contractors want to grow, they usually grow horizontally. Each type of work or division is a vertical market. It is often difficult to grow vertically because it is hard to gain more market share of the same type of work in a particular area or city. Because of this, many contractors find it easier to add another division and grow another vertical market. The adding of a vertical market causes them to grow horizontally as vertical markets are added. During an economic downturn, the trend is for contractors to evaluate these different vertical markets and eliminate any that are not as profitable as they should be or are unprofitable.

When the economy is good, some divisions are so very profitable that they subsidize the ones that are not doing as well. However, because the profit of the whole company is still good, many contractors continue to give unprofitable divisions a chance to succeed. But when the economy has a downturn, the profitable divisions become less profitable and the company does not have the money it once had to subsidize unprofitable divisions. During these downturns, contractors usually eliminate those divisions.

In order to make such decisions, a contractor must have good information to evaluate the financial viability of each division. Divisions or profit centers are best defined as some type of business that the company is going to conduct which is entirely different than any other type of business they conduct. For example, if they are in construction and go into maintenance, this is a different type of business. If they go into retail or a wholesale nursery which grows plants, those are different types of businesses. If a contractor does landscaping and goes into putting in irrigation systems, this is not another type of business because it is still construction. If a contractor does maintenance and also does snow plowing in the winter, this is not another type of business because it is simply another form of a service (maintenance) business.

In deciding what costs to put into each division, it becomes very apparent in the area of the income and cost of sales portion of the divisionalized profit and loss statements. However, it becomes more difficult when it comes to the proper allocation of overhead. The following are two ways in which this is generally handled by most accountants.

The first way is to have a separate division for overhead. Each division does not have an overhead attached to it and just shows a gross profit after the cost of sales are subtracted from the income. The overhead division will show the total overhead for the entire company and it is then applied to each division, usually by a percentage of the sales of each division. For example, if a division has 28 percent of the total sales of the company, it will have 28 percent of the total company overhead applied to it. If that division's gross profit is 31 percent, then it would show a 3 percent profit after the allocation of its share of company overhead. However, this method is not the recommended method as this type of allocation is found to be inaccurate.

Sometimes a division, such as maintenance, will have a lower percentage of sales than construction does (because there is very little material involved in maintenance), but has an equal or greater overhead than construction. This may make maintenance look profitable and construction unprofitable when the exact opposite of this scenario is what is truly happening.

The recommended method is to have each division show its own overhead based on the month-to-month allocation of those costs to each division by the accountant. There are certain costs in overhead which are definitely attributable to a division. For those items that are not so definite, the accountant can use the same principle of allocating those costs as was explained in the section *Budgeting Overhead* (page 465) using the A, B, C method of allocating overhead. This method will give a contractor an instant and more accurate reading as to the profitability of each division.

Another important indicator on a financial statement is the different percentages of costs against sales. A contractor should study his financials to see what percentages labor, material, equipment, subcontractors and overhead are of the total income. These percentages will show a trend of normality over several months of evaluation. For example, a contractor may note that labor usually is between 19 and 21 percent of the income. They may also note that material is between 30 and 32 percent of income. These percentages will vary from division to division, so these ranges should be established for each division.

If, in a given month, the percentage of labor to income goes up, the percentage of material to income should go down because the type of work they did had cheaper material on it, and the work was more labor intense. If the material to income percentage did not go down, a loss will appear on the bottom line and the contractor can identify the reason as a labor cost problem. This can mean the the field was not efficient in producing the work or, as is often the case, there was too much overtime in the time period, which raised the labor costs without raising the income in proportion to those extra costs. This gives the contractor the ability to make the necessary changes which will make him profitable.

If a contractor will follow these concepts and then become a student of his financial statements, he can become an invaluable tool in running his company.

True Life Stories From The Real World

✦ *I Bought A Losing Company*

George (not his real name) has been involved with the business world since he was a child. His father owned his own real estate company, so he became familiar with running a company at a very young age.

During high school he began spending his summers working for various landscape companies throughout the city, eagerly gathering as much information as possible. After developing a love for the landscape industry, he decided to make it a career.

George attended the local college in order to obtain a degree in horticulture. Immediately after his graduation, he began working for a landscape company which had been family owned and operated. Not being a member of the family put a slight strain in the working environment, especially since he had been placed in the position of landscaping supervisor.

This was to be an extremely exciting move for George. The company had previously only pursued irrigation installations and lawn maintenance, but under his control was now going to venture into landscape installation as well. George knew it would not be an easy task, but he looked forward to the challenge. Of course, as he states, he had other motives as well. From the day he started, he had plans of being the owner of the company!

It was a slow tedious process. George sat and watched for over two years as the company slowly regressed. Profit had become nonexistent throughout the company. It was a known fact that the company's reputation was dwindling, as well as the respect of the employees. After all, who would want to work for such a "losing" operation.

But George was not to be daunted. In July of 1988 he knew it was time to make his move - and he did. It was then he purchased 50% of the company, and became a partner with the son of the previous owner. It was not what he called an "ideal situation."

His respect for the previous owner was tremendous, but George knew there would be a lot of things to change.

When purchasing his percentage of the company, George agreed to let the previous owner stay on with a five year contract. He admits it has caused no problems at this point, but the contract hasn't expired yet either.

So now George was the proud half-owner of a landscape company and everything was coming up roses. George was going to turn the company around and make thousands of dollars and take extensive vacations to exotic ports, right? WRONG!

One week after purchasing the company, a contractor's nightmare came true. While unloading bark from a dump truck, the driver realized he could not close the door of the truck. He had a man continue to unload the material from the back of the truck, while he went to pull it forward. But during the blur of seconds ahead, just as the driver got into the truck, it began to roll backwards, pinning the man who was unloading the material between two trucks and causing severe injury.

Most contractors would be worried about insurance rates skyrocketing at this point, but George was mortified. It was only then he found out there was no insurance on the vehicle. A lawsuit was filed against him, which to this day has not been settled.

And this was only the beginning of the troubles. When George had set up the arrangements to purchase his portion of the company, he conducted a review of the accounting books to insure everything was in order. He had hired an outside accountant to check over the figures and make sure everything was correct. Unfortunately, AFTER buying the company, he had found many errors. And to his surprise, he soon realized there were over $20,000 in accounts payable which he was going to have to pay. He stated it was a very trying time for him. He could have declared the contract invalid in the grounds on incorrect information, but instead opted to go through with the deal. After all, he was only 25 years old and felt it was an opportunity very few people would ever see. He already had working crews, equipment, office space and staff, an inventory and a client list. He was taking over an "already made" business.

It didn't take long for George to realize how difficult running a business is from the other side of the fence. According to George, "It is extremely hard to suddenly have to sit in a chair and answer to everyone else. Before, all I had to do was answer to the owner and make sure the work was completed in the best possible way. But suddenly, I was in the driver's seat and I had to be the one to tell the bankers and suppliers and employees that money was tight or non-existent." It was a long and winding road that seemed to never end.

After further review of financial records, George soon realized that in all actuality, he did not have single asset to his name. All the equipment was owned by the bank and the bank was beginning to put the pressure on for its money. George decided it was time to re-structure the company and expand the operation. He continued to install irrigation systems and do lawn maintenance. He also decided to more ardently pursue the landscape construction portion of the business, which had been minimal at this point.

Ever so slowly, business began to improve and he was able to start paying some of the huge debt owed by the company, but it was an extremely hard process. During the years when the company had been failing, the respect of the employees had diminished to nothing. Many did not want to work for the company for fear the company might go bankrupt. What George considered to be the worst item was the terrible reputation which the company had in the local community. They would get occasional calls for work, but when they were asked to provide references, it became a very short call. As George states, "the landscape business is built on reputation in the community and word of mouth. You can advertise in every newspaper, magazine and phone book. All of this will do you no good if you can not tell a prospective customer, "yes, we did this project for Mr. Smith over at XYZ street." My company was to the point where many people didn't want us doing the work because we had a bad reputation with the suppliers and bankers, and our employees were disgruntled.

Yet once again, George was not going to be daunted by all of this. He knew deep down that he really did have a good company; it was just going to be a matter of time. A real problem he had to deal with was ongoing resentment from the previous owner. George's back was against the wall because they had signed the five year contract. So he decided instead of arguing and causing more problems, he would pacify the previous owner by making him feel he was in control. He would go to him with problems which had already been solved or which he already knew how to solve, yet would present the problem anyway. The previous owner soon felt he was still much in control, so the resentment soon faded. According to George they now have a very good working relationship.

George is also quick to admit although there have been extensive problems since he took control, the business has also provided a few bright spots. When he was asked to name them, it didn't take long, but his answer was very enthusiastic. George said that the brightest spot of his "business-owning career" was the day he saw his financial statements with black ink. He had gotten so used to seeing red ink, he says jokingly, "I didn't even know the color black existed on paper."

He finally did it after two and a half years. Just recently the company for the first time since his purchase, showed a profit, a company which has been in business for 41 years! They have been able to purchase new equipment, have replaced almost the entire work force, and have become computerized. George says the real thing which made the company profitable was snow plowing. It was an area which had never been considered, let alone ventured into. But being a risk taker, George decided it would at least provide some cash flow, and it has! In just over a year and a half, the snow plowing operations have gone from $25,000 per year to last year's intake of $75,000. It was a risk that George says he would probably never take again, unless he was in an extremely stable financial situation.

George will also quickly admit he still has a lot to learn about business. He had felt after watching his father run a successful business for years, and having the experience and college education he possesses in landscaping, this would be a fairly simple task. He found out quickly how simple it isn't when you have to make the decisions, and answer all the questions, and prepare a budget and sign the checks, etc... Yet he wouldn't change a thing. The only thing he would possibly do differently is to not have a partner, to do it all on his own.

✦ We Didn't Quit When We Could Have

Jerry walked into his dorm at Clemson University and ran right into his future. Within hours his future partner would visit his dorm room. At first impression they thought little of each other. But from this meeting, a feeling of mutual respect and friendship would develop which has grown to this day.

Both Jerry and Thom went on to get their degrees in Parks and Recreation management, but it was the Nixon Era and they graduated just in time to see civil service jobs cut. At the same time, they saw their slim chance of getting a job suitable to their education. So Thom did what any single, young, fun-loving college graduate would do; he went to Europe to ski, play and work with his brother. Jerry went to Vermont to do the same. But after a short year, reality and responsibility began to settle in on the two young men. Thom went to South Carolina and began to work for a landscape contractor. He wrote Jerry, who also came to South Carolina, and with $1,500 each, they started their company.

For the first two years, the young bachelors ran their business out of Thom's Mom & Dad's house. Mom was the secretary, they were the only employees, and management was their only business. Soon, they decided to expand the business into construction.

Neither Thom nor Jerry are businessmen, they are designers and horticulturists. What followed would be disastrous. The company became proficient at losing money. Year after year the problem would continue until the company was on the verge of bankruptcy.

What they needed was someone with cash and good business ability. So, they got someone with cash but no more business ability than they possessed. The cash gained Dodge (the new kid on the block), a one-third ownership of a sinking ship which had a chance of being salvaged.

I met Dodge and Jerry at a seminar and later at a workshop. They had managed to accumulate $250,000 in negative assets and losses. I was shocked they were still able to operate, and without very tolerant suppliers they would not have. Every supplier had made the company a cash-only account, and was trying to collect on its past-due accounts.

But these guys wanted to succeed and they didn't want to quit. So, what the company has come to call "the blood bath of 19xx" became its only hope. I sent Thom back to drastically cut their overhead while increasing their sales. I must admit, after the workshop I did not expect to see them again. But I did, the next year, at another workshop. When they left for the next workshop, everyone wondered who was going out the door next.

This time we reduced their overhead a little more and fine tuned their numbers. Soon after, they computerized their estimating which allowed them to bid more work without increasing their sales force.

I am personally very proud of them because in four years they survived a large hurricane and turned the company from a $250,000 negative to a plus $100,000 net worth. This proves many times it is better to hang in and fight than to give in to bankruptcy.

However, at this time there was another problem that existed within the company. Jerry and Dodge were finding out they came from two different worlds. Jerry grew up without any siblings in the home. Dodge grew up in a home with six others, and had to fight for everything he got. That made Jerry a non-confronter and Dodge a confronter. The difference in their personalities was so prominent that Thom was constantly mediating between the two. They were a corporation acting like a partnership.

At our advice, they formed an Advisory Board of Directors. This board has not only come up with some good insights on how to run the company, they have become the mediators between Jerry and Dodge. This has eliminated a potentially dangerous situation.

As they look back, they are pleased with many things they have done. At the top of the list is the fact they did not quit when they could have. Even though Chapter 11 may work for some, they feel they are personally and professionally much better off because they did not declare bankruptcy.

They are also glad they spent money that at the moment they didn't seem to have, on professional help. The only regret they did not do it sooner. They thought their answer was more money and a partner. However, more money without the solution to the original problems will only mean more losses. They are quick to say consultants are not any smarter than they are, consultants just have the opportunity to see things the owners would never see themselves.

They are also very pleased with the concept of an Advisory Board of Directors. Already the board is helping them with the legal matters as well as obtaining better and more creative financing. The Board is also helping them keep capital expenditures in line with their financial health.

Their company stands as a model for getting advice, doing what is advised and sticking it out until the bad times turn good.

✦ I Mixed Tragedy and Business and Survived

Jeff is not a quitter. We all have reasons to quit: business reversals, people who don't pay us for our work, or employees who don't do their jobs and cost us our company. Jeff's problems were all of these and more.

He started his business as a sole proprietor working out of a VW. After receiving an agriculture degree from Penn State University, he did what most people in this business have done. He went to work for someone else for a year and a half. He started in small residential contracting and has never fully left this area of work.

But, an event would happen in the life of Jeff and his young wife which would have caused many other small businessmen to quit. Their first baby was born three months premature, and weighed only one and a half pounds. Christopher would go through fourteen operations and spend two years in hospitals and incubations before his parents could take him home for the first time. Christopher's medical problems could fill volumes of medical journals, but like his parents, he was not and never has been a quitter.

For two years Jeff would deal with his problem the best he could. Every day he would spend all his time either working or at the hospital. Time off, relaxation and/or hobbies became nonexistent in his life. His company began to grow giving him even more reasons to practically work himself to death.

Overwork had another negative impact on Jeff's life. Because he lost himself in his work when he was not at the hospital, his marriage began to suffer. To compound his problems, Jeff had surrounded himself with family members within the company. His mother was the bookkeeper, his father worked on weekends and his brother had quit his job as a chemistry teacher to work full time for the company. At no time was Jeff very far from all his problems.

Jeff soon realized things had to change. He also realized certain key players were not going to work for the future of the company. Some of those key players were close family members. Jeff found himself doing a very difficult thing in asking them to leave the company. He then replaced these people through the hiring process very successfully. He also computerized his office and has found this to be a great time saver in producing bids as well as tracking.

Today, Jeff looks back on all these events with the wisdom sometimes only experience can give you. He passes some on to us.

Of the utmost importance, never use your business as an escape mechanism. The balancing of one's time, between work, family and recreation can be a very difficult thing to accomplish. Many entrepreneurs call work their hobby and claim they are building the company for their family. These weak justifications allow them to excuse themselves for a very unbalanced work schedule. For some, this behavior can actually cost them their health and, eventually, the company.

Always find another area of interest in which you and your wife and family can participate together. For Jeff, the realization came as he and his wife became actively involved in national organizations for the handicapped. They began to realize for Christopher to live as fully as he could, they would have to work together to help him. Today, Christopher walks, talks and goes to a special school where he is learning. These are all things the doctors originally said would never happen.

Whether it be your church, a club or organization, or some mutual hobby, an entrepreneur who wants a successful home life, must get involved with the family in something outside the business.

Jeff now knows how difficult it can be to have close family members working for your company. So often, the safest place to get employees is within the family structure. You know these people, you trust these people, and you feel they will be loyal to you and your company. Many times, all these benefits work for you. However, on occasion, the job or the company is larger than the abilities of a family member and then you are left with a major problem. Most contractors procrastinate entirely too long before doing anything about it and then it is harder and more expensive to do. Jeff lost thousands of dollars on bad jobs because he waited too long before taking action.

And last but not least, get help for yourself as quickly as possible. Jeff admits to his lack of business understanding. He has now come to realize the importance of seminars, workshops, trade conventions and consultants. He credits a tremendous turnaround in his company to this type of continuing education.

Jeff has a goal - to build his company and eventually, put a manager in place to run it. He wants this not for himself, but so his son Christopher will never become dependent on the state to care for him when Jeff and his wife can't. We hope he will succeed in this important goal.

Growing and Staffing Your Business

Growing and Staffing
Your Business

If you plan to increase the size of your business, there are several concepts which may help you plan to cause your success and avoid some of the common pitfalls associated with growth.

As you read this section, I believe you will discover these are not "Cloud 9," abstract ideas. Everything included in this section is based on experiences gained by running and observing successful companies. I'm going to share with you some of the experiences of my friend Howard Eckel. Howard's philosophies helped him move up from a starting position of district manager in his company (with only $70,000 annual sales) to General Manager of his company (with over 3,000 employees in 34 states and $80 million in sales).

While Howard's experience is in the field known as the *Green Industry*, these concepts will work successfully in every aspect of every business, regardless of the industry to which they belong. They are, in fact, important features in a company's growth, and its ability to staff for growth and exceptional profit improvement.

The size of your company is not important, nor does it give undue advantage in the pursuit of success. How you and your company are perceived in the marketplace and having the adequate staffing to serve this market are the key components to success. Staffing and how you function in the market are closely intertwined. Acquiring labor and personnel problems are ultimately part of the marketing function. Establishing the right marketing position enables you to generate adequate sales volume and this is the first step in solving the "labor problem."

Everyone hires people. The trick is to keep them! This section will first address how marketing and sales techniques can be used to solve the staffing problems by generating quality sales. It will then provide some more insights which directly concern hiring and managing people.

You'll be studying the importance of positioning and marketing, as well as reviewing the 80/20 rule. There will be some things you might have to unlearn. Some right and some wrong answers to sales and sales strategies will be examined, and you will learn the reason for the success of the smartest plumber in Tramansburgh, NY. You'll learn how persistence pays off, if the pot is filled with chicken or feathers, and a little about PMP (to find out what that means, read on).

You'll check out using clipboards as brains and what to do with all those business cards. You'll also learn about purple elephants with pink spots and how to effectively use door hangers.

By now, you've figured out this IS NOT an "all you have to do is" section. Instead, you'll discover what Howard did and what you can do. You will learn how to turn small accomplishments and innovative ideas into actions which help <u>cause success</u>.

Success Is Spelled Position

The experts in the retail business say the key to success in their business is location, location, location. In the construction industry, the key to success is market position, market position, market position. Position is everything. In essence it is: "There is no competition, I am the competition."

Position is attained by reputation.

Reputation: good name, esteem in which a person is held, credibility, character, of good repute.

If you are a builder, plumbing or electrical contractor, masonry or concrete contractor or contractor servicing the utility industry, reputation in the market place is position.

Go to the telephone yellow pages, turn to the section where your business is listed and count the firms which are in the same section competing against you. If you're not careful you can become discouraged. There are so many in relation to the available market. Some names drop out each year, but always new ones jump into the market. More names are added to the section every year.

As you scan the list of competitors, you know only a few are at the top, quite a few are clustered in the middle and a lot are drifting in and out of the bottom ranks.

The 80% - 20% Rule

The 80% - 20% rule states that 80% of the goods and services are produced and sold by the top (or best) 20% of the providers and that the other 20% of the business activity is produced and moved by the remaining 80%.

If you analyze your competitors in the yellow pages, the 80% - 20% rule starts to assume validity. While quantities may have differed, it is fair to say that everyone started out with equipment, knowledge and people. Then why do some succeed, some drift and some fold?

The top twenty percent have one common denominator: an excellent reputation in the market place.

An excellent reputation in the market place is not achieved by having the resources to afford a quarter to half page advertisement in the Yellow Pages. It is not quite that simple, yet it is not difficult or complicated. It certainly should not be a function of time, money, or equipment. Reputation is the result of satisfied customers.

Everyone can purchase the same lumber, bricks, concrete, wiring or tree for resale, the same truck to deliver it to the job site, people to build it, and charge about the same for the work, perhaps more, perhaps less.

So how do the twenty percent do it? By developing a reputation. They determine what they do best and the market wants their services. Then they serve this market and serve it well! They end up being able to make more sales with fewer salespeople, and produce more with fewer people and less equipment. They have less office staff and paper work. They seem to have more cash to work with, more time to think ahead, to plan. They are always out in front.

Does Servicing the Market Work?

This question can best be answered by sharing my own experience. I started with the company in 1963 as a district manager of the Eastern Massachusetts operations. I took over what could be described as a disaster. Through one circumstance or another there had been three different managers in the territory in three years. The work force consisted of five production people. When I met them the first morning, two were still drunk from the weekend. Now I was down to three! The answering service was receiving approximately two calls every quarter. Territory sales volume was $70,000 per year. Big business right?

To top it off, even though I was working for a tree service company, I did not know about tree surgery. I'd grown up in the nursery, landscape design and construction business. I knew what tree surgery was from college, but I had no experience in pricing and selling it. But I was in the service business. And service is service! I'd clean your attic if you wanted!

In five years, we had annual sales of approximately $600,000. I was manager and salesperson, we had one supervisor, and the company and I were making money.

Developing a reputation in the market place by servicing the market did it!

Then I switched jobs within the company and was responsible for expanding our utility services in the Southeast. Sales went from three million annually to nine million in three years.

Developing a reputation in the market place by servicing the market did it!

In 1979 I was put in charge of all company operations (over 120 separate operations in 40 states and the province of Ontario, Canada). Combined sales volume was fifty two million dollars. In six years, by the end of 1985, sales were one hundred and twenty nine million. Profits had increased more than threefold.

Developing a reputation in the market place by servicing the market did it!

If you think I accomplished these achievements by myself, I have a bridge to sell you! I worked with a lot of different talents and expertise, and I learned as I went along. I never stopped learning.

Learning how to service the market place, any market place, is not difficult. It should not be any mystery. A strong focus, a little learning, a little unlearning, and a reputation and success can be yours.

Some Things to Unlearn

If it is all so simple and uncomplicated why does the 80%-20% rule holds validity? In part, one of the reasons success remains elusive to some is due to what we have been taught.

We all were brought up in an education system which taught us if numbers are added or subtracted, we arrive at a total. This total is called the answer. Deductive reasoning provides the answer. An answer which was previously not known to us is deduced.

If we arrived at the correct answer it was great. We moved to the head of the class. If we arrived at the incorrect answer, things were not so good, and we were made to go back and keep working the problem until we deduced the correct answer. We tried until we mastered the process of achieving the correct answer. It didn't matter how long it took.

We leave school, take a job and eventually start a business. We soon learn whatever we do in business, whatever actions we take or don't take can be quantified. Numbers are developed to measure what we have done (or have not done). These numbers representing action or lack of action are added up and a total (answer) is arrived at. Except now, in business, it is not called the answer, now it is called "The Bottom Line." The process, however, centers on deductive reasoning.

If the answer is not what we wanted or hoped for, it's back to the drawing board. We try again (and again, etc.). Thank goodness for the concept of "hope springs eternal." Except in business, time is a concern. Time is money!

Usually our first professional contact outside our field is the bookkeeper or accountant. They are the professional scorekeepers of our actions. They professionally reinforce and perpetuate what we were taught. They add up all the numbers we created and arrive at an answer, *the bottom line.* It is still deductive reasoning.

One analogy is football. You play the game. Someone will keep score, the number of downs, and all the statistics. The scorekeeper can't help you go forward, he can only tell you how far you've gone. He can't tell you if you've won until after you've won. Cross the goal line more times than the other team by following the "Playbook," through a series of well executed plays, with the will to win. And you win!

I believe the way we are taught early in life has a great influence on the degree of our success. The 20 percent were taught right, or reversed some of the things they were taught to get it right. I find there is one common denominator among those who are the most successful. They know exactly what they are doing and what their actions are going to produce in quantitative numbers. Their actions are deliberate and produce numbers which add up to the answers they want. This is inductive reasoning.

Because they know what they are doing and what they will accomplish, they also benefit from economy of motion. Their energies are rarely spent on false endeavors, going down blind alleys to dead ends.

I have worked with people who instinctively knew what the bottom line would look like if they took a specific action or took a certain direction. I was not born with natural skill, I had to acquire it.

I soon learned if I wanted to succeed, I had to figure out where I wanted to go, wanted to be, and then take the necessary action which would enable me to get there. I knew the answer I wanted, so I started to work on the problem.

I had an accountant friend who figured out I was determining the answer and then working the problem. The thought process was so alien to him, to the way he had been trained that he was almost accusative when he stopped me in the hall one day and said, "You use inductive reasoning!" I smiled and said it was true. I knew the goal or position I wanted the corporation to achieve. It was the answer. I simply figured out the actions we would take to generate the numbers that would add up to the answer. I simply worked the problem. He still appeared frustrated, since it was not what he was taught. It was inductive.

The difference between the 20 percent and the 80 percent is deliberate acts versus random acts. The 80 percent create random numbers by random, often unconnected acts, which when totaled, usually are somewhat of a surprise. The 20 percent reverse the process. They know the total they want, and once the answer is known it is their process to work the problem. Efforts and actions are tailored to create numbers which give the answer they want. It is inductive.

The concept "know the answer, work the problem," is similar to "create the solution and let others react."

I am forever grateful when early in my career I learned to reverse the deductive process I'd been taught, and started to use inductive reasoning.

Some Right and Wrong Answers

The concept I apply to most phases of the Construction Industry, and I have already stressed it, is that we are in the service business. We don't sell products as much as we sell ideas, people, equipment and knowledge. When you approach it from the service angle, from the "what do people want" angle? everything becomes quite simple. I am not a complex or particularly bright man, so this approach has appeal for me. We don't have to worry about product life, market saturation or all the other marketing phraseology. People and other firms want service and they always will. Don't fight it; give the customers what they want. It is the answer.

My career sales statistics were mostly developed from cost-plus or negotiated-contract or time-and-materials jobs, with some choice lump-sum bidding. I learned early I could not be all things to all people, and there was business which I preferred my competition to have. Using this approach I backed into the "position" which I think is the answer.

"I'm in business for myself, I am my own boss, I can do anything I want to do." Welcome to the eighty percent!

"I am in business to satisfy the needs and desires of others." Welcome to the twenty percent!

How many of you have, or use, a General Electric computer? No one has them. They never survived. In the fifties and early sixties, General Electric, which has a reputation as being one of the best managed manufacturing companies, spent millions of dollars developing a computer and attendant software. In those days the company which developed the hardware also developed the software. There were no separate software companies as there are today. Millions of dollars later, GE closed out the computer business. No one wanted their software so no one would purchase their computers. The software was what the GE programmers of software thought the consumers should have, not what the consumer wanted.

In the meantime IBM did everything. They made the hardware and the software. Software that people wanted. IBM software is still in use today. The criterion for other hardware and software is still "IBM compatible."

How many of you ship and receive goods via the railroad? Not many. Most of them went bankrupt. Railroads would ship your goods to where you wanted them, but it was usually when it suited them. They could never tell you where the shipment was while it was enroute or when it would arrive.

I've shipped via the commercial trucking industry. They'll tell you when it will arrive, even the route it will take if you are interested. You can call while the item is enroute and they can tell you where it is. Is that why the commercial trucking industry has grown and the railroads move bulk cargo and turn their rail beds into bicycle paths?

Experience convinces me finding the right niche or market position, then servicing this market segment, is the component the successful firms in the industry have in common. It is one of the answers.

Successful firms in the construction business do not think of themselves being in the concrete business, the electrical business, or the maintenance business. These are the services they offer. They organize their knowledge, equipment, people and inventory to serve the customer. All actions are focused on making the customer feel important. This is one of the answers all successful firms have in common.

Once you decide you are in the service business, success is close to being a foregone conclusion. Service will require more work and perhaps the need to make more actions than you first thought, but the actions are not complicated or difficult. Determine who you are, what you do best. Then determine whom you want to serve and serve them.

I used to tell salespeople just starting out once they were on a prospect's property, they should come out with a sale. The key was finding out what the client wanted, then structuring knowledge, equipment, inventory and people to fill the customer's wants. Literally, a no-brainer. What they had to do was listen. Then if they wanted to be really successful in sales, they would determine the client's satisfaction upon completion of the work. Further, they would continue to service the client without being asked. If you retain anything from this book, this last sentence is the key.

The Smartest Plumber
in Trumansburg, New York

A great deal of the volume increase in New England, the Southeast and the corporation, came about because of "the smartest plumber in Trumansburg, New York." I am not an effective cold caller. In fact, I am not even a good salesman. This realization was what helped me the most. I listened to anyone and observed everyone. I also had just finished five years working for one of the smartest nurserymen-landscape contractors in the business.

I learned a lot during my five years at White Nurseries. I learned we charged just about twice as much per hour as the competition. We always had more work than they did. I learned there are some prospective customers or businesses you are better off letting the other guy have. I learned there are no short cuts. I learned about letting your customers sell work for you.

What has a plumber in upstate New York got to do with all of this?

Trumansburg is a little village about twenty minutes northwest of Ithaca, New York. My wife and I, with three small children, lived there when I worked for White Nurseries. I went to work one morning as the landscape designer/salesman and Mr. White called me into his office. I'd been employed there about six months. He asked me what I'd purchased lately. It was early in the morning and I really wasn't focused. I hadn't the faintest idea what he was talking about. He kept pressing me as to what I'd purchased lately. I finally remembered we had the local plumber replace the hot water heater.

The three children were in diapers, disposable hadn't been invented yet and diaper service wasn't available in a village of

1700 people. So, I told him I'd purchased a hot water heater. He asked me who installed it. I was getting a little agitated by this time and came up with the sarcastic remark, "I'd also paid for it." I simply couldn't figure out where he was going with this conversation, and I really thought it was none of his business. It had been crisis time when the old one let go. He told me to calm down and then asked me what I would think if the plumber had called my wife back after a few weeks and asked if the recovery cycle was fast enough for all the washing she had to do?

I thought for just a second and responded, "I'd think he was the smartest plumber in town." With that, he slid a clipboard across the desk at me. It contained a list of all the customers we had done business with in the last year. He told me to visit every customer, answer any questions, make suggestions, drink coffee with them, and show them we were interested in them.

He told me there were only 50,000 people in the market place and we couldn't afford to lose any of them as customers. He told me when customers, properly treated, would sell more work for us over a bridge table or on a golf course than he and I combined ever could. He called it "smart buyer one-up-manship."

His theory, and it later proved to be correct in almost any business anywhere, is people will brag to their peers about what good shoppers they are, how smart they are, to do business with so and so. "We just don't worry about the grounds any more after Virgil found White Nurseries." We did not advertise it or make slogans. We serviced the customers' property without their requesting it. A smart plumber!

Mr. White positioned himself and his firm as the premier in his field in the market. People felt pampered and bragged to their friends about the service they received. He was a twenty-percenter.

Whenever managers say they could do more business if they had another salesperson, I always wonder if they have their customers enrolled as salesmen. It certainly is less expensive than more salaries, benefits, commissions and automobiles.

Mr. Eckel, Are You Here Again?

I've mentioned what I walked into when I started out in Boston. I didn't know much more than the theory of tree surgery, and the company's reputation was in tatters in the area.

The division manager came up from Connecticut and said he'd handle the sixteen customer complaints left over from the previous district manager. Now I hated complaints, but I didn't have much else to do so I asked to be the one to settle them. I reassured him I wouldn't give the company away. He was glad to be out from under them. I thought I could salvage something. At least I had sixteen calls to make. We were averaging two calls on the answering service every quarter. Looking back, it would have been easier to start a new business under a new name. Remember, I've said I wasn't too bright.

In the meantime the home office sent me the old client list. All 110 names! With sixteen complaints, two telephone calls every quarter, and 110 old clients, golly, hang on Virgil, here we go. I settled the complaints with my tail between my legs; I think I managed to salvage three of them as clients. They had festered for far too long.

I am not an effective cold-call salesman. Never was and never will be. So I haunted those 110 old customers. You talk about service. A leaf didn't fall that I did not know about it. I was prowling around one place in Cambridge, Massachusetts one morning; I had not yet knocked on the door. The lady of the house called out and asked, "Mr. Eckel, are you here again?" I told her I didn't have any place else to go. She invited me in to have a cup of coffee with her and the houseman/chauffeur.

So what's the point about all this? Do I want the reader to feel sorry for me? Don't! The territory sales base was $70,000 the year before I arrived. My approach didn't set the world on fire, but we managed to do $90,000 my first full year. The second year we did $120,000. By the third year we hit almost $240,000. The fourth year and beyond we were over $364,000. The answering service was handling 12 to 15 calls a day; in fact, I opened up a second number and service. No cold calls either. Those satisfied customers were out there on the golf greens and at the bridge tables bragging about how smart they were and what great service they were receiving from me and my company.

The Soup Pot is Filled
With Either Chicken or Feathers

Years ago, when kitchens had wood stoves for cooking, it was not uncommon to have a big, old, cast iron pot on top of the stove. Water was added as needed along with leftovers. It made for some great, if not varied, soup.

A very old Virginia gentleman told me years ago he came from a county where they invented poverty. He said that their soup pot either had chicken in it during the good times or feathers when times were tough. Sounds a lot like the construction industry. Either more work than we can handle, more things to do than time, or more production capacity and people than orders.

Clients and prospects wander around the grounds on the first sunny day every spring and then your phone falls off the wall. Everyone wants it done at once. It's called Spring Fever! Doesn't take long to build up a work backlog. The Spring Fever phenomena helped me the first few years in Boston. The established firms would have a four to six week backlog. Out of frustration, people called me. You talk about service. I'd sell it in the morning, meet the crew for lunch, give them the order and they would do the job in the afternoon. Brilliant market strategy on my part!

I would get right back to the new client whose work we just did. Check everything over and make sure they were happy. Often I would be there giving the crew the next order while they were cleaning and loading up. Talk about client satisfaction and personalized service. We were the best. Clients began to spread the word. Business picked up.

Within two years I was facing about the same situation everyone else does. Everyone wants it done now. We had a larger client base and started to have a potential backlog problem. New prospects and clients all wanted work done now! Even when we built up to more than seven crews and business was booming, I established a policy a new client order had to be worked within two weeks. There might have been a few times we didn't make it, but we tried to adhere to the ten days with a vengeance. I did not want to lose them to someone else who was hungry and could do work that afternoon.

It was necessary to take command of the workload. I wanted chicken in the pot twelve months of the year. The staff did also.

First, if a client called me, I apologized. I should have been servicing them, bringing things that needed attention to their attention. After I'd finished beating myself with birch whips over the client calling me, there are some things I knew about that client.

1. They were satisfied with work previously performed.

2. They had confidence in my organization and me.

3. They considered me the expert.

As the expert, take command of the workload. What you recommend is what they will do. Use this confidence to your advantage.

If new plantings are desired, you're going to have to do it when it is best for the plants, not your schedule. That same thinking can work for you when timing is not a critical element. "It would be best for the tree, the shrubs, whatever, if the pruning was done in August." How can they argue with that? "One of the best times to feed the trees would be from late August through the late fall." We offer a winter discount on pruning. Chances are you'll end up with two foremen doubled up and working together. Also, without leaves, there is about a twenty-five percent less bulk to clean up." "I'll have the crew stop by within a day or two and re-move that large broken limb. I'd then prefer to hold off and do the rest of the pruning when everything has been out in leaf, say late June or early July." "We can plant the flowering shrubs within ten days. I do think it best we wait until all the new growth is finished on the evergreens and other ornamentals before prun-ing. I'll schedule pruning for late June, early July."

You are the expert. Use your technical knowledge to postpone some tasks to non-peak times.

It must be done ethically, but there are many tasks which can be accomplished other than in the peak season.

You begin to level the sharp sales peaks and valleys to a more un-dulating line, and all of a sudden you have a handle on the man-power problem. It is much easier to build a work force in the off season. You have work when other firms don't. You select and hire when others are not competing for work.

In the spring the "Help Wanted" advertisements fill the classi-fieds. It is not the best time to be trying to build the work force. Build the work force when the other firms are not hiring. A more even workload allows you to do this. We would hire in the late fall and late winter. It gave us a chance to become acquainted and absorb the new employee. I'll talk about this in a later sec-tion but we didn't hire people, we absorbed them. I did not want to be like Pogo in the comic strip, when he said: "we have cap-tured the enemy and we is them."

Leveling the workload goes a long way toward solving the labor problem. I took great delight in writing the corporate personnel department every spring and saying we did not need any assis-tance. We had enough people to meet our needs. In fact, the last two winters I was in Boston, we kept men busy from other adja-cent company territories who had no winter work.

My customers and I sold enough work that when I transferred out of the territory in the late fall of 1968, there were enough orders on file to keep the seven crews busy until the middle of March, 1969.

P.M.P.

I was first introduced to P.M.P. at White Nurseries, Meclenburg and Ithaca, New York. It stands for Preventative Maintenance Patrol. Sounds corny but clients loved it. It organizes and formalizes taking control of the workload. I have uses P.M.P. in the landscape and tree care businesses with equal success. Once you explain to a client how P.M.P. works, the usual reaction is "You mean I don't have to concern myself any more, you'll take care of things? Great!"

P.M.P. works best on repeat customers. Unless asked, I never mentioned it to a prospect or new client. When they began to demonstrate confidence in my organization and me by repeat business, I would then bring it up. It was handled low key, not promoted, or specifically advertised; it was offered as an additional service, usually while talking to the client on the property. P.M.P. tied in with taking control of the workload.

I've seen two gasoline service station operators in different parts of the country use the P.M.P. concept. Both became millionaires. I explained the P.M.P. concept to an automobile salesman. He applied the concept to his activity. He achieved great success and he was selling AMC Gremlins at the time!

When I received that magnificent client list of 110 names with their sales history, the same sales volume pattern was evident in the caring of trees which I'd experienced with the care of ornamentals at White Nursery. For many and varied reasons people would put off pruning for three to five year cycles. I'd point out this pattern. Chances were they took care of pruning and feeding requirements when the need was very obvious.

My pitch was simple; highs and lows in tree and shrub care were not good for the trees and planting. Starvation, then gluttony was not good. Plantings need consistent care. If a program was laid out and care given when it was best for the plant, less money would ultimately be spent because the client would not be playing catch up or doing corrective work, and eliminate the large expenditure every three to five years. The plantings benefited the most from consistent, timely attention.

P.M.P. was nothing more than a property inventory and schedule which listed by month specific tasks that needed to be done for each plant species on the particular property. I always made the point I was not selling labor but rather technical knowledge performed by professionals. They could hire warm bodies to pull weeds or cut the grass. We supplied the technical aspect. It always seemed to make a great deal of sense to the client. I'd even lay out the lawn care program if the pickup truck bunch wasn't doing it properly.

Find a niche and position yourself as the professional, the expert. You're creating the situation and others are reacting.

I made a rough sketch and property layout with all plant material located and listed. Back at the office I would make out all the individual, monthly work orders which the property inventory required. Lots of work but once done you had a year's worth of orders in the file for that client. With enough clients, all of a sudden you have a lot of orders for each month of the year.

Forsythia is thinned and fed in early May, after blooming. (Nice rainy day jobs.) Rhododendrons were scheduled for mid- to late-June for feeding, flower pod removal, and touch-up pruning. Major renovations were scheduled for non-peak times. Tree pruning was scheduled for late summer or winter.

In the meantime, I concentrated my activity. Our client base was not spread all over creation. With most of our work coming from referrals, the client base tended to be concentrated in pockets. You might only have a half-hour's work of lilac pruning on a single property, but there were other jobs close by so half or full crew days could be scheduled.

The production people were fully involved in the concept of P.M.P.; they knew we were out to please and satisfy the clients. They all knew P.M.P. was part of the plan to insure year round work. You don't think a foreman wouldn't stop by for fifteen minutes worth of Forsythia thinning on the way home from a tiring day, when he knew there was winter work with that client?

It was teamwork at it's finest. People with somewhat different goals working together.

Clip Boards

I transferred a bright young foreman to an eastern city to take over a territory after the manager who had been there for three years passed away.

We expected some reduction in sales at first until he gained sales experience, but the territory went into a real slump. We were unsettled, as we knew technically he was tops. I went to the territory to spend some time with him and to boost his flagging spirits.

The first call we went on was a referral. The lady wanted a tree in the front yard removed. I'm not one for tree removals, but we did have an experienced work force who needed the work. Located under the tree limbs were mature Rhododendrons and an ancient wrought iron fence.

My young, technically-experienced friend studied the situation for a few minutes, then threw a price. I was standing back, but I could see the lady's face fall. She said she'd talk it over with her husband and get back to us. We had been recommended, but prior to the recommendation she had secured another quote that was about $100 less.

When we were back in the car I told the younger manager that his knowledge of the business was working against him. He had left the prospect with no alternative but to go with the low bid.

I explained, unlike him I did not grow up in the tree business or work as a foreman prior to selling, so I had to learn the time study. His price was fair; however, due to his experience, he had arrived at the price in his head. Not many homeowners were experienced in the tree removal as he was. He needed to do a little education.

I told him that due to my lack of time study experience, I'd have a clipboard with me and in front of the prospect I'd write out the following:

1. Load crane truck with sheets of 4 x 4
    and 8 x 4 Plyscore for "Teepeeing" the
    Rhododendrons and the wrought iron fence.
    (Note: We always had Plyscore on hand
    just for these situations.)

    Hrs:_____

2. Bring reinforced mortar box.
    (Note: We had a 3 x 5 mason mortar
    box with angle iron welded around the
    edges. It was a land barge. It could be
    winched across a turf area, heavily
    loaded, and never leave a rut. It was
    also good for lifting men and materials
    up steep slopes.)

    Hrs:_____

3. Drive to work site.   Hrs:_____

4. Set up plywood tents over plantings
    and fence.   Hrs:_____

5. Remove tree in small 18' sections.
    Lower each section.   Hrs:_____

6. Use mortar box and crane winch
    to slide debris across lawn areas and
    load on to trucks.   Hrs:_____

7. Clean up thoroughly.   Hrs:_____

8. Haul to dump.   Hrs:_____

9. Drive to yard, unload tools and
   equipment.   Hrs:_____

                                          Total Hrs:_____

Crew Rate (Composite Rate) times total hrs = $_____

                                          Dump Fees: $_____

What the young man could do in his head because of experience, I had to list and time, step by step, on a yellow pad on a clipboard. I'd give that rough, handwritten time study sheet to the prospect along with the formal quotation form. Now she would have more information than just a price when she discussed the quote with her husband.

Even if I didn't get the job, pity the other fellow who would show up and fail to do it exactly as I outlined! I made the lady an expert on tree removal. Even the slightest scratch on the wrought iron fence or a broken twig on a Rhododendron would make for one of those days when the contractor should have stayed in bed. Sales soon improved in the territory.

Shirt Pockets

I also asked my young manager where his 3" X 5" index cards were? Didn't he carry them in his shirt pocket?

The odds were he was not going to get the tree removal job. They would go with the low price. Also that the odds were 50/50 there would be some damage, and the prospect would be thinking "If only I'd gone with…."

Every time you are on a property it is an opportunity for future sales, even if you lost out the first time. The 3" X 5" index card is indispensable. It is stiff and firm enough to write on while walking around. It fits conveniently in the shirt pocket.

Take a quick glance at the other key plantings, make note of them. What about those Rhododendrons once the tree is removed. Perhaps several new trees could be replacements or perhaps transplant them to another shady spot. What about a feeding and insect control program? Every property has multi-sales potential.

I scan the property even while walking back to the car. I'd make note of key plantings and trees. I took the information I wrote down on the index cards and used it in future mailings. We did what I call fill in letters. They had an opening and closing paragraph preprinted with whatever service of function, say feeding, we wanted to promote. A personalized paragraph was also included (See Example 1 on page 409).

As the client, "Am I impressed? I didn't give them the removal job last spring, much to my sorrow, yet they send me a personalized letter alerting me to the Maple by the patio should be fed. The manager, in long hand, at the bottom of the letter gave me additional suggestions on caring for the Rhododendrons. No one has taken an interest like this in a long time. I must tell the Virgils and the Jones what great service I am getting and I haven't even done business with them yet."

Position yourself in the market and then service that position. Develop a reputation by showing an interest in every contact made. All of a sudden, you have more people talking about you, pre-selling you and your firm, generating more potential sales than you could imagine.

All Those Business Cards

Just about everyone in business has a box of business cards in their desks and perhaps even a few in their wallet or purse. Usually the cards are replaced because the information on them becomes outdated. The telephone number is changed, the address, the title, something happens which renders the cards worthless. It's very unusual for them to be reordered as is. The original remainder of the original box sits and gathers dust.

It is no different in a large corporation, particularly a company which is growing fast. Cards are thrown away as obsolete long before they are used up. I was not surprised I received an inquiry from the home office, wondering about "all those business cards"? I was using over two boxes a year . . . over a thousand cards. What in the world was I doing with them? No one ever reordered, let alone reordered consistently.

When I first started out I had little use for them. Business cards were for first-time contacts; goodness knows I certainly had few of those. I always tried to leave a quotation sheet with the prospect. That had the company name and local telephone number on it so I didn't use many cards.

As business grew I started to use them with a vengeance. I did not use them much in initial contact, still stayed with the quotation form. I used them to continually keep my presence and that of my company in front of the client.

Thank goodness for that smart plumber in Trumansburg, New York. He taught me to keep the clients always in mind, and then the client would keep us always in his mind.

Business is usually concentrated in certain geographical sections. If you're servicing your clients you will be referred and your prospects will usually be near your client base. It does not take much time to stop at a client's property while driving between appointments.

It does not take much time out of even the busiest appointment calendars to stop and quickly survey a client's property. You are looking for major problems so it won't take that long.

You haven't the time to stop and talk, so don't announce yourself, don't ring the doorbell, just leave your business card wedged in the doorjamb or near the mailbox.

How would you feel if you returned home and the arborist or landscape firm that did some work two months ago had left a card with:

> *"Stopped by, everything OK.*
> *– can't spend any more of your money!"*

> *or*

> *"The Oak by the garage may need to be fed,*
> *I'll keep my eye on it for a few months."*

> *or*

> *"The new Rhododendron and Azalea planting*
> *should be soaked more. I'll keep checking them."*

Would you feel important? Next time when talking to a friend about almost anything green and growing, would you mention the fantastic service you are receiving from so and so?

It is so simple and so easy. Yet most of the salespeople I discussed this with would not have any idea what I was discussing or come up with several reasons why they just didn't have the time between appointments. "I could never keep track of everything. What if they are home and wanted to talk?"

If you are spotted and the client wants to talk, just tell them that you are already late for another appointment but you just had to check the tree, the plantings, …whatever and could we make an appointment for ____ later on? "Oh, Virgil! Even though he was late he still thought enough of us to stop by."

I can remember properties and specific plantings from years ago. Even if I did not, I have those handy 3" X 5" cards in my shirt pocket which can be placed directly into a call up file when I arrive back at the office.

I admit it is not very glamorous, just detail and nitty gritty, one $28.00 order at a time. But with my clients selling for me, I did not need another salesperson to help me keep seven crews busy year around.

I do admit as business grows you sure go through a lot of business cards.

Door Hangers

White Nurseries of Ithaca and Mecklenburg, New York used door hangers, not to obtain new business but to insure customer satisfaction. This outfit was fanatical when it came to making sure the client was happy. Remember there were not many people in the market and they needed every potential prospect.

Mr. White developed a simple door hanger which the foreman left when the job was finished. Often the client would not be home when the crew completed the work. Every client was called by telephone within twenty-four hours of a job completion, but as the salesman/designer, I would always make an appointment to go over the completed job with him or her in person. We wanted to let the client know when the crew arrived, provide a condensed description of what was done, and when the crew left. A door hanger with a detachable return card filled the bill. There was also a spot on the card for the foreman to make suggestions about things he noticed and a place for him to sign. (See Example #2 on page 411)

Then on a detachable postage-paid return card, there were five boxes the client could check and a place for comments, signature and a place for the client's telephone number. It was a great idea. There was no way we were going to have an unhappy client. We had all the bases covered. Wrong! It turned out that the foremen hated to leave them and often did not. The reason was the five boxes that the client could check off on the return postage paid card:

❑ Crew did a great job.
❑ Stop by and have a cup of coffee any time.
❑ See me about what the foreman noticed.
❑ See us next year.
❑ I am mad!

Human nature being what it is, the foreman did not want to open the floodgates to criticism. After all, who does? It took a lot of coaching before they left the cards on a regular, consistent basis. At first they would only leave the card if the client had been right there while the job was done. They knew they were fairly safe if the client had not complained during the job. What turned the resistance around, however, was when a few of the cards came back with "Crew did a great job." These were posted on a bulletin board and the competition was on between the foremen to see who could get the most compliments.

Most door hangers are used to solicit business. White Nurseries used them to insure customer satisfaction which ultimately led to referral business. No mention was ever made of "Tell your friends." The smart buyer syndrome would take care of it.

Another benefit was derived from all this prompt attention by follow up telephone calls, door hangers and personal visits: People paid these bills faster. With all this attention there was no reason not to.

Elephants Painted Purple With Pink Spots

Everyone wants service and attention. Everyone!

I mentioned previously I had been involved in utility line clearing sales, and we went from three million in sales to nine million in three years.

The line clearing business is large volume with very, very low profit margins. Why it is so competitive, with every one willing to work for almost no profit, was always a mystery to me.

Our company handled the annual contract negotiations on our cost-plus accounts with personnel from our home office. At one point in my career, this became my job. I would literally spend hours arguing and negotiating for a nickel an hour more in chain saw billing rate on a two million dollar account. Five cents per hour isn't much, but times 3 saws per crew, times 35 crews, times 1850 crew hours per year, it could almost make or break an account's profitability.

We soon learned having someone from the home office going out once a year to negotiate billing rates was not going to get the job done. As one client pointed out, whenever he saw a representative from our company home office, our man was asking for more money.

It really was a no-brainer. We set up a schedule to call on each division and even some district utility offices each quarter. Four visits a year with only one of those visits concerned with money. On the other visits we asked, "How are we doing? How could we do it better? Were you satisfied with the service?"

On my first visit to a utility division office to ask these questions, the utility division manager pulled a U.S. Army 45 pistol out of his desk drawer, cocked it and then laid it on his desk, muzzle towards me. I couldn't even tell if the clip was in it or not!

Until this time, he had not said a word to our division manager or me. After he put the gun on the desk, he asked me if he had my attention.

I am not sure if my voice broke or not, but I assured him he most certainly had my complete and total attention. He proceeded to lecture me on rendering service to the customer. He was paying for service, and what he wanted, he wanted. If he wanted elephants painted pink with purple spots to move logs on a right-of-way, it was our job as the contractor to provide them, not to question or procrastinate.

We went to lunch and had an amicable talk. "Talk" means I listened.

What soon became apparent was our local supervisor had taken to heart our constant rhetoric about the need for cost control on these low-profit utility accounts.

The utility manager might ask for another truck or a specialized piece of equipment and our supervisor translated the request into more cost for our company. He would stonewall the request.

We took our supervisor aside that afternoon and explained what we really meant about cost control. We gave him some highlights on the improvements to the bottom line we might expect as a result of additional sales over fixed costs. We spent most of the time explaining our new approach of servicing the client at every level.

We soon scheduled a meeting with all of our supervisors. The main topic was servicing the customer, how each of us was key and each of us had a part to perform and what it was. It worked. We soon set the standards which the customer expected all contractors to meet.

When you start positioning yourself in the market, make sure everyone in your organization, even if it is only two other people, understands exactly what you are trying to accomplish, how everyone plays a part, the part they play, and the benefits that everyone can anticipate.

If ever there was reinforcement to the service concept, that utility manager was it. Sales went from three million to nine million in three years.

On Being the Best Thing That Ever Happened to a Landscape Architect

If ever there was the ultimate, perfect situation for an adversarial relationship, it is in the bidding of landscape projects designed by and supervised by landscape architects.

Talk about gladiators, duelists, arch foes, love/hate relationships. We have it all when the landscape contractor and the landscape architect pair up, or is it square off?

Both parties are working at cross-purposes. One, the architect wants the perfect job. Not a scratch on the tree trunk, not a needle off the evergreens and definitely no broken twigs on the deciduous material. After all, as the representative of the owner and the creator, the architect probably selected each plant and tree.

The landscape contractor was awarded the bid in the majority of the cases because he submitted the low bid. Time is money; production is crucial. The contractor has not the time to sit or stand around and contemplate the intellectual aspects of the project. He wants it completed, accepted and his money paid. He needs to move on to the next project. Besides the "Pickup Truck" crowd were bidders, so the bid is pared to the bone. It is a low bid in every sense of the word.

As I [Charles Vander Kooi] have said, "There was no Ding-Dong factored into the contractors bid." You could spell "Ding-Dong," as "patience."

If you stand back and look at the situation and the potential for strife, it reads like a daytime soap opera on TV. It won't get any better until you, the contractor, break out of the arena and begin to approach it differently. Why is it the contractor has to change?

Why can't the other guy, the architect change? Whoever said life was fair?

I'd been brought up in the protected atmosphere of the landscape designer/contractor. The design plans were drawn in-house and executed by in-house personnel. Usually I had been the designer and then the supervisor on the job. I'm going to fight with myself?

So what a shock when I went to New England with the large corporation. Their primary business was arboriculture. However, depending on the local manager's experience, landscape contracting could be undertaken in a territory. In fact, the company had a commercial landscape department which did nothing but bid and execute large landscape projects in the U.S. and Eastern Canada. These projects were always designed under the auspices of landscape architectural firms. Therefore, it was company policy no one in the company should compete against the landscape architects by doing design work. We were to remain strictly contractors.

Being a diligent new employee, I did no design work. Instead I began to call on the better landscape architectural firms in Eastern Massachusetts. When I called on these prestigious firms to introduce myself and have the company's name placed on the list of reputable bidders, I received two basic responses: "Who?" and "We have all the qualified bidders we need."

The only thing left for me was bidding publicly funded projects, advertised to the world in various local papers. Talk about the pickup truck crowd. You never saw so many pickups until you went to a public pre-bid meeting in Eastern New England.

I still remember walking into my first pre-bid meeting on a school project. A sea of faces all turning toward me, scowling. You could just read their minds: "Another idiot. Wonder if he will bid stupid? Where do all these jerks come from?"

I was in shock. I have never seen so many sharpened pencils in my life. These guys must have come from five states. I did not turn in a bid, not for months. I'd go to most of the pre-bids, paying my ten or twenty dollars for a set of plans and specs. I'd check the site and specifications but couldn't screw up enough courage to bid against that sea of faces. I was petrified I'd be low.

I can't remember just what triggered my attention at one of the public openings, but the architect or the architect's representative always seemed to be nervous until a certain point while the bids were being opened and posted. Sometimes their nervousness dissolved early in the proceedings, other times the nervousness dissolved further along in the procedure.

It finally hit me. They were worried until a bid came in below the budget. Collectively, that sea of faces was occasionally smart enough not to bid low to stay within the owner's budget if the architect's design was too expensive.

My perception was clinched at one bid opening on a school project when everyone came in above the school board's published budget. The architect had egg on his face with the board. His design was too expensive.

In general, I thought about the architects and the problems that they face. I put the adversarial aspect of my thinking aside. I put myself in the architect's shoes. As an in-house designer, I had a slight feel for some of the architect's problems.

The Shoes of the Landscape Architect

The Landscape Architect primarily makes a livelihood by designing and subsequently supervising the installation of the design.

Reputation is the key to a successful career.

Reputation is established by pleasing designs fostering satisfied clients. Staying within the owner's budget is part of the satisfaction scene. The design, properly installed, must also flourish and prosper to create that satisfaction.

The Landscape Architect, due to the time constraints imposed by field supervision and quality control, can only undertake a certain number of projects during a given time frame. Thus, to a certain extent, the architect's income is capped.

The Landscape Architect is held responsible for bids being within budget. Thus, he must be up-to-date on influences affecting bids, availability of material and installation costs. The architect is receptive to any action that will solve or reduce the problems he faces in:

A. Bids within budgets;

B. Quality installations; and

C. Advice about landscape maintenance as an enhancement to his finished project.

I had lived my business life so far as a fanatic on serving the client. Based on the creed of service, I felt I had the key to the architect's heart. I was going to be a hero, the cavalry with bugles blowing, to save the wagon train. I was going to offer service to

the architect that would help him. Increase his stature in the eyes
of his clients. Allow him to improve his income.

Servicing the Architect
– Bids Within Budgets

The concept is simple.

The project is pre-bid by someone who is knowledgeable and has timely information on material availability. They create credibility by ultimately not submitting a formal bid on the particular project.

I was prepared to pre-bid public openings, low bid projects valued under $20,000 to $25,000. These projects pay the rent for the architect, but don't hold much glamour for them. They dislike spending inordinate amounts of time or effort on each projects, yet cannot afford to have the bids come in over budget.

The idea is to service the landscape architects and gain their confidence in you and your firm. You may even want to make yourself available to scout material via telephone or catalogues.

I concentrated on two medium-sized firms, where success had begun to put strains and time constraints on the staff. The principal architects had a staff turning out designs. They would do public-bid projects. To help pay the overhead. Their staff was not always experienced. Due to time pressures their work was not always carefully reviewed by a principal.

My sales pitch was simple. I was willing to offer material scouting services to them. I'd review plant lists and comment on availability. I would alert them if a particular plant item was in tight supply or unavailable locally. If Pinus Stobus 12'-14' were not available locally and had to be brought in from another state, a contractor had to add a large freight bill on top of the plant cost. The project might exceed budget. If the size were dropped down to 8' - 10', the material was available locally. It was a lot of work, a lot of weary night hours and long distance telephone calls tracking material.

They became confident in me. I grew in stature in their eyes. I wore a high profile. I was making their life just a little easier. My pitch was simple. I was an experienced contractor who knew the business. I'd grown up in it. I was there to make their life a little easier.

Why go to all this effort? I did not want to be another face in a sea of faces at public pre-bid meetings. I wanted in on the selected contractor bid list. A known group of four or five reputable, quality contractors to bid against. The bids would then have no more than five percent spread between all of them.

I wanted time and material work. Architects are always faced with small, nuisance jobs. Big money clients always have a little pet project they want the architect to do. If an architect can turn it over to someone he trusts, someone who doesn't require a lot of supervision, someone he has implicit faith in, namely you, he can get on with the projects that are more meaningful.

If you get yourself on a selected bidders list with a few architects, then use the information in the section "Troubleshooting the Estimating and Bid Process" found elsewhere in this book, you can be very profitable working with architects.

Quality Installation

Once you do obtain one of those preferred jobs, whether through bid or negotiation, keep the service profile high. Remember you are trying to establish a reputation as a quality contractor. Once on a project, go out of your way to communicate with the architect.

A. Leave verbal progress reports, daily, with his secretary or key staff member.

B. Notify in advance when to expect deliveries on the site or when a major undertaking is happening.

C. If you have a question, consider whether you can solve it at the architect's office rather than calling the architect out to the site.

D. After the project has been completed and accepted, and the guarantee period passed, continue to make brief suggestions to the architect about ways to improve the maintenance.

If the architect perceives your efforts are on his behalf, make his life easier, more profitable, enhance his stature with the client, and help him make more money, he will be prejudiced toward you every time.

One last word: Stay away from the great big, multifaceted jobs. Let someone else tie himself in knots. There is more profit in a few small to medium jobs than there is in the big ones. Of the thousands of contractors who have entered into bidding the larger projects, name me five that have survived over the last forty years. I think what happens is success. They make money on small to medium jobs, and then think they are unstoppable. Before they know what happens, they are history. Conversely, I can name you five firms in the service end of the green industry, all of which have been around sixty to eighty years, and collectively do over a billion dollars of sales annually.

Position yourself in the market, then service that position.

Staffing - For Better Not Worse

"If I only had the people, I could do more." It is the most common lament. Regardless of the size of our business, people or the lack of them - production capacity if you will - is the single biggest constraint according to most firms in the Green Industry.

It is not going to be better by itself.

I just read that the "baby boomers" have matured. There will be 8.5% fewer 18 to 24 years olds entering the work place during the next four years. I recently talked to the director of the school of horticulture, the State University of New York's Agricultural and Technical Institute at Alfred, New York. They have two-year programs in landscape maintenance, construction, interiorscape, greenhouse and nursery management. He said that the average graduate, with an associate degree in one of these disciplines would receive 15 job offers. Not interviews, offers!

Having sufficient, dependable and productive people has always been a problem. Now it is going to become even more difficult. You are competing against every industry for people. Your ability to attract and keep good people is essential; your future success will depend on it.

In previous pages I have stressed you are in the service business. These services may be the tree business, the landscape business, the lawn care business or the maintenance business, but people perform them. What you are really in, then, is the people business. The product you sell in the service business is people. How well your staff performs directly impacts your success. The ability to attract and keep additional good people also affects your future. You can have all the nursery stock, the latest equipment, computers, fancy software programs, accountants, attorneys, widgets and gadgets, but if you don't have people, good people, you are dead in the water.

You wanted to make money, be your own boss, plant and care for trees and all things green. You wanted to construct, to irrigate, to grow. No one really told you about the people part of this. You knew you had to have them. You perhaps just didn't know part was going to be a hassle, take so much time, be so frustrating.

Everyone hones their sales skills and management techniques, obtains the latest computers and software, and keeps up with the latest technical innovations. However, very few people spend time and reflective thought on honing their ability to manage people. If the ability to manage people is not a natural aptitude, you will continually have people problems, unless you acquire the aptitude.

If you run into a technical problem, the state university extension service or the county agent can be contacted. The suppliers will give you detailed descriptions and instructions on how to use their products. The truck breaks down, or a hydraulic system lets go on a tractor and you call the mechanic. Who do you call to solve your staffing problems? There are not too many places you can turn to for help and advice. You and only you can come up with the remedy, the solution.

Your trade association will provide training aids, safety manuals, even sample policy manuals and procedures. I can give you the name and number of a top flight firm which will train any of your people in the required skills for landscape maintenance, construction and tree climbing. In some cases, they will even come up with the people to train.

Yet staffing is such an issue, such a problem.

Articles on the staffing problem have been written and read by just about everyone in the industry. About six months ago I listened to a tape which touched on some highlights for solving the industry's labor problems. Neither the articles nor the tape are run away best sellers. People are just not moving ahead of their labor problems after reading or listening. Why? I don't even know you, but I am fairly certain the problem is you.

The very skills, drive, determination and desire which drove you to be successful work against you when it comes to people and how to handle them.

I've worked too long with too many managers not to be confident when I say you are probably the problem. Just don't take offense. It is not a crime, not a sin and certainly not a lack of intelligence on your part. In fact, it is probably a result of your drive and intelligence.

The problem is you're doing what you wanted to do, started out to do, and enjoy doing. You wanted to go into business for yourself for many reasons. I have never heard a person say they went into business, any business, just so they could hire and manage people. Therein lies the root of the problem.

Let us try some inductive reasoning to solve the problem. The answer we want is "a competent cadre of people at every level so we may be successful in what we started to do in the first place."

The Second 80/20 Rule

There is a second 80/20 rule. I came up with it the hard way, by making mistakes. The second 80/20 rule says: "that sometime in your career, you will pass a critical point. That critical point is when you can maintain a good sales volume with only 20% effort. The other 80% effort should now be spent in managing the business. Particularly, the people end of it."

When you started out, sales were critical so you spent all your time obtaining them. Slowly your reputation built. Soon your customers were selling work for you. It is at about this point you should start to spend more and more time in managing what is happening. I'm not talking about office systems or fancy computer programs. I'm talking about managing people. I am talking about really getting into the people aspect of your business. Instead of spending time on sales which will come in anyway, or studying up on all the fun equipment there is to play with, start thinking about people.

Human nature being what it is, some of you who read this book are going to find what I write and advise will not help solve your staffing problem because you will not change. You are not mentally prepared to clear two hurdles:

1. You are in the service business. You are in the business of selling and managing people.

2. You, and only you, can and must master the solution, the answer to the staffing problem.

All the lectures, how-to tapes, magazine articles or books, will not help if you do not accept and endorse these two concepts. These are two of the necessary steps needed to arrive at the answer.

Working the problem for staffing is not quite as easy to accomplish as is knowing the answer and working the problem for a sales or marketing situation. There is an important key difference: in the market place where you sell, the market place itself creates and sets the atmosphere. The successful solution to your staffing problem depends on your creating the atmospheres in which a staffing plan can function.

The Two Atmospheres

You have taken a positive step toward solving your staffing problems by developing a reputation and some leveling out of sale volume. This gives you one of the two atmospheres needed to solve the problem. You can attract prospective employees' attention by advertising for people when others can't. That is a tremendous help. Setting yourself up as the premier firm in the right market position creates one of the atmospheres needed to solve the staffing problem.

There is one more atmosphere needed. I can define it, but only you can create it. It is easy to define, yet difficult to create if you do not have the natural ability to effectively manage people or have not acquired that ability.

There are a lot of reasons any firm has staffing problems, but if you stand back and are objective, there is only one key, one real answer, and only you can provide it. Did you ever hear the expression: "A great idea at the wrong time." The "wrong time," is really having the wrong atmosphere or no atmosphere for the idea to thrive in.

You can develop the concept, plan to make all the right moves, and take all the correct actions, with a marketing concept, and it won't work if the marketplace doesn't have a need. It won't sell. The same is true in developing a solution to a staffing problem. You had better come up with something people want, or it won't work.

What Do People Want?

I have been involved in managing just about every type of business in the green industry in just about every major city in the United States and Ontario, Canada. I have heard it all when it comes to staffing. "We have a unique situation here. This town is different. I simply can't pay enough to compete for labor with other businesses. Our unemployment rate is down under 3% and there is no one to hire. You simply don't understand how it is here, et cetera."

You won't convince me. There is an answer and there are actions you can take to solve the problem, but you must change the way you think about business and people. You, the manager, owner, entrepreneur, major domo, whatever you call yourself, can solve the problem. In fact, you are the only one who can.

There are statistics accumulated over the years as to what production people, foremen and supervisors ranked as most important to them and most satisfying. Money, wages if you will, ranked consistently fourth or fifth in importance. The priorities have never changed through the years.

Opportunities for advancement, recognition, and working conditions, always rank above money in the surveys. They confirm these rankings in business school. Give me a job, put a roof over my head, fill me with food and take care of me when I'm sick, and I'll be happy. Wrong!

I read a while back where two thousand people a day left East Germany and headed west even after the wall had been down for several months. They had low cost housing, food, medical care, secure jobs and yet they left in droves. Why? They wanted an identity, they wanted a future, they wanted to be in charge of their own destinies.

What happened in East Germany substantiates the statistics. Wages alone will not solve the labor problem. People want more than just the food and housing that wages provide. They want a future, and they want a pleasant atmosphere to work in. I did not say easy work, just a pleasant atmosphere while they do even the most arduous tasks. They want to be recognized for their efforts. They want to be able to see where their efforts will take them in the future. They want to be involved in, as well as be part of, the enthusiasm.

If you want to solve the staffing problem, you must address these human needs and wants, and you'll have to do it sincerely. It is very difficult to change, and it is impossible to be something you are not. If you pretend to create an atmosphere, be totally enthralled with people, but you are not, people will see right through you. You must change your ways and your attitude and the attitude of the entire firm, whether it is two or two thousand people.

You are going to develop a "Corporate Culture."

Corporate Culture

You need to develop a culture which is going to center on and focus on people. You are going to get as positively excited about people as you would if you had landed the most profitable ongoing contract in the world, or just purchased the newest widget on the market.

Corporate Culture is spelled atmosphere.

It is difficult to explain a "corporate culture." It is people being emphatic, fair, consistent, communicative, encouraging, concerned and thoughtful. Managers with few, if any, staffing problems usually have one common denominator, the Golden Rule, "do unto others as you would have done unto you."

Too many managers, owners, entrepreneurs follow the other two golden rules:

1. I've got all the gold, so I rule.

2. I haven't got all the gold yet, so get out of my way.

I used to fly out of Cleveland almost every Monday morning. I always kidded that I liked to fly Northwest Airlines, and would except they were always on strike. Their ticket counter always seemed to be closed due to labor difficulty.

To quote excerpts from a news story about Northwest I just read:

"Union leaders accustomed to fighting with management now sing its praises. Out on the front lines our employees are bubbling with enthusiasm. 'People are smiling,' says a customer....They've repaid more than 1.3 billion dollars of debt more than two years ahead of schedule....More importantly they are overhauling a corporate culture famous for autocratic, penny-pinching management into one which puts people - customers and employees-first. That the CEO Checchie and his close knit team may be the right people to turn things around as they learned the importance of serving the customer working for...."

The article went on to say the turnaround, the new corporate culture, is the result of actions three new managers took. In less than a year, these three people initiated a change in the corporate culture which affected 42,030 employees plus hundreds of thousands of customers.

The act of genuinely caring, being emphatic about customers and employees, creates an atmosphere which filters through the entire organization. Regardless of the size of your firm, if you can come across as being people oriented and, in fact, are your staffing plans will be effective because there is an effective atmosphere for them to function in.

You will also find out because you and your employees are in accord, additional sales come in. Following the "Other 80/20 Rule" did not hurt your volume at all. The market place would much rather do business with, come in contact with, an organization that is enthusiastic and cheerful.

We Captured the Enemy and We is Them

It has been said, "You are what you eat." "You are what you hire" is even more real. The staff which goes out daily to produce your sales is really the company. It is not only a case of how they perform the work, but how the customer perceives them.

The crew goes on the job site. Do they represent you and your philosophy, or are they free floating, projecting their own individual image? There are legitimate arguments for either hiring the inexperienced and training them yourself, or hiring experienced people immediately capable of producing. I really don't think it makes any difference as long as the new hire, regardless of experience, is absorbed into your organization, and it is not a case of you and your firm being absorbed and becoming a composite of what you hired.

I admit I am a fanatic on the so-called corporate culture. To have a firm represented by a staff which does not reflect its firm's creed and philosophy is chaos, and a quick way to lose control. You have worked hard and long. It would be a shame if people working for you do not represent you, do not understand what you want your firm to be. I don't care if you have just two others working for you. They need to be extensions of you, your ideas and your competence.

There is a way to accomplish this without smothering your staff's individuality. In fact, you can develop a firm or corporate culture that everyone buys into and, simultaneously, harnesses the employees' individuality to mutually benefit you and them.

How you and your firm appear to the world is up to you. Creating a corporate culture takes time, the focusing of your attention and energy, and a great deal of sincere enthusiasm. The effort will be worth it. Just ask the management of Northwest Airlines.

This all sounds well and good, but where are these people coming from?

Not Business Cards Again?

I have said I really don't care if experienced or inexperienced people are hired, it is how they are absorbed into an organization and then how they project the organization's image which is important. It is all well and good to say I don't care where they come from, but I also know it is a problem at times just to get people to interview.

Let's say you advertise in the off season. Yes, you are going to have people show up who have been laid off (the expendables from your competition). So what? There could be a diamond in the rough among them. It is worth a try.

Advertising in the off season begins to add to this corporate culture. I am still employed but my boss has cut everything to the bone and here is the other outfit always looking for people in the off season. Rumor has it they are okay to work for. I think I'd better check them out.

Believe me, the thought process does work like this and you really can attract some competent people who are looking for the long pull and not the 10 cents an hour more. If all contractors continued to hire only each other's people by whatever method or timing, we eventually would run out of people. New people also need to be attracted into the industry for the future.

I learned how to do this from our utility operations. All this growth, going from 3 million in sales to 9 million during the three-year period I referred to, did not come from taking over some other contractor's crews. These were new, previously unbudgeted people which the utility division needed. It required additional new people.

The rapid expansion went off without too many hitches. There certainly are no textbooks I am aware of which tell you how to do something like this, so how is it done? It was done with those darn business cards again.

As I traveled around, I asked a lot of questions. Where do all the people come from? I found out they came from every other industry and situation imaginable.

The account manager had trained his supervisors to always be on the lookout for good people. When I asked him just what he meant by good people, he proceeded to give me one of the greatest pep talks I'd ever heard on the merits for working in the business and in particular, our company. Talk about enthusiasm!

Every time he or one of his supervisors came into contact with an upbeat person in any situation, he would go out of his way to talk to that individual about employment with us. They did it enthusiastically.

I guess I'd seen it happen in the past, but it never sank in as to what they were doing. Remember, I never said I was very bright. I remember pulling into a small town service station with the account manager one day. It was before self-service was common. A neat and personable young man came out and very efficiently waited on us. The young lad was a pleasure to behold as he briskly went about a rather mundane job. He brought enthusiasm to a boring task.

When our manager got out of the car to pay the fellow, he gave him his business card. He complimented the attendant on the enthusiastic job. "If you ever consider a change in careers, please call me. I work for a company that has unlimited futures for people who are not afraid to work and be happy while doing it. We may not have an opening right away, but fill out an employment application. We are always expanding and looking for people who want to make a future for themselves."

Do you think this young fellow threw that business card away? Our manager probably gave him the first pat on the back he'd received in months. When we pulled away from the pumps, his feet weren't even touching the ground. I asked the manager if this was standard practice. He said that he and his supervisors did it routinely. Later on, I asked a supervisor about it. He said he had applications on file for his area and never had trouble filling crew requirements. He said, in fact, he'd just given his business card to a waitress who waited on him at lunch. We were on the lookout for everyone who seemed to have potential. There are women filling almost every production, sales, administrative and management job in the company.

We are competing for people with every industry. If you have something to offer people, don't keep it under a barrel. Start selling!

These fellows would build a relationship with the high school counselor. It doesn't take much time to stop in and introduce yourself and to keep the contact viable by a periodic stop or a quick note. Oh! Be sure and leave some of those business cards. The manager and his supervisors would go to all of the local churches, all denominations and all races. Look up the minister, priest, pastor, whomever, and find out who the bright young parishioners were. They had futures to sell and they sold them! They would contact these young people, and send them employment applications and return envelopes if they could not see them personally.

It is a lot like having your customers sell work for you. If you have a guidance counselor and a minister or priest or rabbi spreading the word, you'll have people lined up outside your door. If you wait around to see what the advertisement in the newspaper brings in, you end up short of people to interview.

I will not spend much time on the interview process. I am a lousy interviewer. I sell the job. They can't wait to sign up when I finish with them. In fact, I sent out a memo that said in effect, "Don't let Eckel interview new prospective employees." I was good at scouting out people already in our employ for advancement; just don't let me hire new ones.

It goes without saying you will check all references!

I asked the account manger if foremen hired. He said no. They might furnish a name, but he and the supervisors were the only ones to hire. He said his experience was many foremen would not suggest anyone with more ability than themselves (foremen). If the company was to meet its growth plan, management had to continually look for people able to fill future needs at every level.

Shave and a Hair Cut

Just about everything you purchase has instructions and operating manuals. Often there is a trouble-shooting section that says if the product doesn't start or even function correctly, try such and such. Sometimes even a telephone WATTS number you can call toll free if you are really stymied.

Did you ever hire anyone who had his or her own personal operating manual? Page after page of instructions on how to obtain peak performance from them. No? There is also the problem of having no extension service, county agent, manufacturer or mechanic to help you with your staffing problems. No wonder every firm says labor is its largest problem. No experts to help, no instructions, and you sure didn't go into business for the sole purpose of managing people. What's a person to do?

First off, you're going to acknowledge "The Other 80%-20% Rule," and you're going to learn where successfully managing people can be just as much fun as making that big sale. But it takes time and thought. It takes enthusiasm. It takes focus.

Someone will occasionally say the employees are assets of your firm, just as inventory and equipment are. Then the statement is usually left hanging. Not much is said about maintaining these people assets. It finally dawned on me; I was going to treat people just like prized assets.

Equipment comes with operating instructions and maintenance manuals which tell you when to grease, change the oil, rotate the tires and turn the widget to the right. All assets need looking after. So you do the same for your people assets. Write a set of operating instructions for your employee assets. If you have another employee or two, obtain their input, their suggestions on what should be put into the employee instruction and maintenance manual.

All those business cards have people lined up outside your doorway. They believed you when you said you worked in an exciting, never boring field and your firm had unlimited future potential for a person wanting to work. They want to know more about it. How does it all fit together and work?

The manager who showed me how to obtain people from all walks of life also ran a great operation. Everyone in uniform, no long hair, no long beards to become entangled in the climbing line, taut line hitches, equipment always looking good. The utility client was always satisfied. A showcase operation.

He and the supervisors who did the hiring followed the same format when interviewing. It was real simple - in the manager's words, "Never hire a person and then tell them to get a shave or haircut."

So simple, yet so important. When you are interviewing people, tell them exactly what you expect before they accept, before they are hired, not after. What you expect of an employee becomes the basis for the operating instructions you are going to write.

A new, inexperienced person should be able to review a step-by-step schedule of items and facts they will have to learn before moving on to the next level, bite-sized steps the new hire should expect to master. A prospective employee should see at interview time what they are expected to learn. They will be expected to learn to identify 10 trees, 10 ornamental deciduous saws and power saw chains, etc. The next three months has a new and expanded list of things to learn and master - CPR, first aid, more plant identification. Whatever service line you are in, it can be broken down to logical progressive steps.

Every three to six months there is another plateau to attain. Nothing is ever boring if you are developing a future for yourself.

You need a complete set of job descriptions and responsibilities for each job function. Sounds like a lot of bureaucratic red tape, but believe me, it isn't. If you don't organize all the arrows so they travel in the same direction, one of the random arrows flying around will get you!

Write complete step-by-step job descriptions and responsibilities for each position. Identify knowledge and proficiency levels to be mastered before moving on to the next level.

Interview prospective employees with a complete development program which contains detailed descriptions of the various job functions within the company. It is an organized approach to the possible career paths available. It is the key to your ability to effectively meet staffing needs. These learning schedules and job descriptions are your operating instructions and your employee asset maintenance manuals.

Establish your own "corporate culture." Capture the enemy and he remains captured. They become you and they are assimilated. I said earlier I didn't care if we'd hired experienced or inexperienced people. It really didn't matter, because we had developed a program which detailed step-by-step what each employee was expected to learn in each job function before moving on to the next job.

Everyone can set his or her own level of development. If you want to move on in the organization, develop your maximum potential, you can. If you want to stop and stay at one level, and some do, you can. Your knowledge and acquired proficiency affects how far you progress and what you earn. Merit raises are earned by your effort, not by whim or interpretation, and definitely not by time in grade.

Equal and The Same Are Not the Same

You have just hired a bright, promising, enthusiastic new person. Who will be his new supervisor? Virgil needs another helper so you place the new hire there.

Who is this guy, Virgil? Does he have a large project going on and needs more help? Is he just short of people? Just what does Virgil do best? Do you really want this enthusiastic new hire turned over to him?

The Declaration of Independence states that all men are created equal. It does not say that all men are created the same! Is Virgil the person you want to train this new, impressionable apprentice?

A couple of things became apparent to me over the years. Not everyone was a trainer or even wanted to be. Second, if hiring to fill a vacancy, always ask yourself: "Am I hiring into a weak or inexperienced situation?"

On one account we had two senior foremen. Both had been with the company over 15 years, both were highly thought of by the customer. One foreman had turned out 12 new foremen during his tenure, the other none.

The one just got a great kick out of taking a ground person who was ready to learn to climb, teaching him all the ropes, literally, and then (polishing him off) so he could take over his own crew. The other foreman just wanted to do a good job and work with the same people every day. He just hated changes in his routine. Still, he was a very productive foreman.

Figure out who your trainers are. Talk to them; tell them that training is going to be part of their job from now on. You might even want to change their compensation package to reflect the additional responsibilities.

If you don't have anyone who is a trainer, guess what? You are it. You will have to become involved. Eighty percent of your time is available. There is nothing more important than your people assets.

Your new hire has a complete outline, program, job description, whatever you want to call it. He or she understands exactly what knowledge is needed to get ahead in the firm. You know what else you accomplished when you spent all of those hours writing job descriptions and specific responsibilities and things to be learned?

You just started to organize the organization.

You just gave Virgil, if he is your training foreman, a complete outline of what he needs to teach. The new hire knows what needs to be learned.

It is the ultimate in communication.

You have spent a great deal of time developing job descriptions, responsibilities and apprentice/new hire learning schedules. You have a foreman can teach and wants to teach. Those to be taught have a clear idea of what they need to learn; those teaching have a copy of the same schedule and know what needs to be taught. You are in great shape!

Go to it, Virgil!

Wait Boss! You are not finished. Virgil has plenty to do, but there is more for you to do.

By the Way,
I Have Been Meaning to Tell You

You are going to monitor everyone's progress!

Quite a few people manage from anger. They let dissatisfaction build up, and then let fly. There is no communication when there is anger. The recipients of the anger go to either a siege mentality or on the offensive.

I was always delighted to learn at raise time, when I was counting on a raise, I wasn't going to get it because of some deficiency in my performance. Great way to generate employee loyalty. Too many employers simply cannot bring themselves to do performance evaluations except when it is time for wage reviews. Hiding behind raise schedules, unfortunately, is more of the norm than it should be.

If you have job descriptions and a "things to learn schedule," a professional yet personal atmosphere is created. It is easy enough to sit down with someone periodically and do a review, a performance evaluation. All too often the review is controlled by the calendar and not by the employee. "I am ready to be reviewed, I think I have learned this phase of the schedule." Or a spontaneous review prompted by the owner or manager asking the employees, especially the new employee, "How are you doing? Are you ready to talk over your progress?"

Remember the statistics about what people think is more important than money. Recognition. Call it attention if you want, but please don't forget me. Notice me, be interested in me, and I will do even better.

At raise time it is far better to be able to say: "You have learned more of the items than I expected you to since our last talk. I think that a merit raise is in order. Here is the section I want you to concentrate on now. What parts are giving you the most trouble? What can we do to help you? Here is where we think you need to spend more time. Virgil is aware of this and will work with you."

Please do not tell me at raise time what I'm not doing right. Keep my progress reports timely. Don't wait until you are frustrated and angry. Tell me promptly so I have a chance to learn – work with me.

If He Had Only Known

Things are going well. The new person is with the training fore-man. They know what they have to do. What about the other people?

For years our company had supervisors and managers do six-month foreman evaluations. There were about thirty items which were rated unsatisfactory, satisfactory, or excellent. The evaluations were faithfully filled out and sent in every six months. They were placed in the employees' service record file. I was aware of this, but had really not thought too much about the process.

I'd been out in Idaho calling on a client and spending some face time with the crews. The home office did appreciate their efforts and sent me 2000 miles to individually tell them. There goes that corporate culture again.

When I arrived back at my desk Monday morning, there was a manager from our West Coast subsidiary. They had been grow-ing by leaps and bounds and he had asked to put out more crews. We were doing about 20 million in sales with the customer and we sure wanted to keep them happy.

He could come up with the foreman and climbers but was hurting for a supervisor. Well, I had just seen a foreman in northern Idaho who was doing a great job for us, was willing to move, and would be perfect. He had 7 years experience as a foreman and was ready to take on supervisory duties.

I went over to the personnel section and checked out his person-nel folder. I was planning to weed out nonessentials and send the pertinent information to the West Coast. In going through the folder, there were 14 "Foreman Evaluations" forms faithfully prepared by various supervisors or managers. For 7 years, for 14 forms, the people rating this foreman gave him excellent to satisfactory in all items but two. There were more excellent ratings now than in the beginning, which made sense, but the same two items were consistently rated unsatisfactory. They weren't important items in the grand scheme of things. To this day I couldn't even tell you what they were. . . sloppy glove compartment or something equally obscure. If he wanted the job, I was sure it was his. Those two items would not stand in his way.

I really became excited. I knew this foreman fairly well. I knew if he had known about those two unsatisfactory items, he'd correct them. He would straighten himself out. He was a prideful man and would work on any deficiencies others might see in him.

That knowledge is what got me excited. I went back over to the personnel office and asked the department head if the foreman ever saw these ratings. I knew the answer before I asked. I guess I was just being ornery. He said, "No, the evaluation sheets had never been seen by the foreman. Just filed away."

It is difficult to explain, but I was so happy. We were a good company and had been for seventy-one years. Now, I knew we could even be better! It was exhilarating.

I sat down with the personnel people and we redesigned the evaluation sheet. Actually, all we did was make room so the foreman could sign his name, signifying that the supervisor had gone over the evaluation with him. Personnel was excited and had the new forms printed quickly and sent them right out to the supervisors and managers with a cover letter.

The cover letter stated that the reviewer was to go over the evaluation with the foreman and have him sign it, signifying he had seen it. We were really going to roll now. The work force would really become even more proficient. Right?…Wrong!

The foreman evaluations came back in. I think that twenty or thirty out of the 600 forms had a foreman's signature. All the unsigned forms had ratings which were very typical and not much different than the ones already in the individual's files - a few "excellent," quite a few "satisfactory," and a few "unsatisfactory" on each form.

Well, back to the drawing board. I never said I was smart.

Once again we sent the newly-created form out to the supervisors, with instructions re-emphasizing the foreman's signature on the form was mandatory. We wrote we wanted them to talk to the foreman in person about the evaluation, go over each item and discuss how each rating was arrived at.

The forms all came back in with the foremen's signatures. Great. Except all the ratings were marked "excellent"! There were a few "satisfactory," but not one "unsatisfactory." The evaluator just wasn't going to confront the foreman and tell him what he was deficient in. Let someone else do it. In order to keep a good working relationship, unless things really went haywire, they were going to let things slide along.

We set up a series of meetings around the country on: "How to do evaluations." We simply had to make this system work. We had to overcome the resistance to face-to-face evaluations. It took some convincing, but we finally sold everyone on the concept of a face-to-face, fair evaluation would be received very well if it were formalized by utilizing the evaluation form. There is something about a piece of paper between the evaluator and the evaluated that impersonalizes the situation. Having a formal process produces respect for the process.

What Did I Learn in the Southwest?

I learned another important aspect of training and evaluations because of a frustrating situation on an account in the Southwest.

We were authorized 24 crews but could never field more than 18. We had a local office, management, supervision, everything. We were losing profits because we could not produce the additional sales over our already established fixed costs. We could never seem to staff the crews we were authorized.

I sat down with the account manager and the four supervisors on the job to see if we could find a solution. Admittedly, the wage scale was low. We had been on the account for years. If it had been a new prospect, we probably would not have tried hard to land the account. It was one of those jobs you'd like to see your competition have. It was a cost-plus account; the customer set the wage scale, and set it low in order to keep their tree trimming costs low. The customer had not learned where people, not money, trim trees.

No sense in spending time on "what should have been." Let's solve the problem. Everyone agreed low wages was the problem. I asked to see their EEO reports. We had to keep these for the government. The reports showed we had, in fact, hired people. We just did not keep them.

I asked the supervisors if they saw each crew each day? Yes, they did. I then asked them if each crew was of equal proficiency and the foremen all at the same level of leadership. Their response was negative. I then asked if certain crews were continually short of people? Was there a correlation between a weak or less-experienced foreman and quitting? The look on their faces I shall never forget. They realized that they were continually hiring into the weak, troublesome crews. There were a few foremen that managed to keep their men and even develop proficiency in them, but the supervisor's trouble was with the weak foreman.

We established a few ground rules for the account.

The supervisors would spend more time with the weakest crews. They did not have to see the stronger foremen every day. Spending their time equally was not going to solve the problem.

New hires would not be placed with the weaker crew.

The supervisor told the weaker foreman that he, the supervisor, was going to spend time with him, to help him attain personal proficiency in his job as foreman. If, after reasonable time, the supervisor felt that the foreman was not going to become proficient, he had to decide whether to discharge him, or demote him and move him to another crew.

The top climber from the stronger crew was moved over as foreman. The supervisor would organize his schedule to spend time with him. We did not want to get in the syndrome of telling someone, "Here are the keys to the truck, you have just been promoted," then watch them drive off into the sunset, thinking they were all alone.

Hey! We didn't work miracles, but nine months from our meeting we had a complement of 24 crews. A happy ending? In three years we were off the account. We decided we had better places to put our people, management and equipment. Running a sweatshop was not in our corporate interest. It did not fit in with our corporate image or culture. The client had an antiquated accounts payable system, and they were slow to pay, so we had more working capital tied up than the industry average. There is some business we would rather the competition had.

I did learn from the experience. All people are not created the same, just equal. Never hire to the weak crew. You can upgrade your work force by organizing your time. People will respond if they think you are there to help them improve. People want to do better, sometimes they just don't know how.

How Come He is Sitting In the Class?
He Could Teach the Course

We were on a roll. We were getting our act together so we could meet the growth plan. We had a handle on training the apprentice/new hire, the climbers and the foremen. We had training schedules, job descriptions, lists of specific job responsibilities. There was no stopping us now!

Except that we had no real program to train the trainers.

The company used to bring its field management into week-long seminars. The agenda covered the entire spectrum we all face in the business: safety, equipment, customer relations, production techniques, technical innovations, equipment maintenance. It looked good on paper and it covered every subject which needed to be covered, but we were getting mixed reviews and results from the program.

We tried to rewrite the training programs. We were not very good at it. If fact, we were terrible. Try writing a training program someday. It is very difficult. I was frustrated; our collective efforts were not filling the bill. We looked at programs from outside the company and they were too general, too broad. Our previous experience with having guest lecturers from the university had not been well received either. It was too theoretical for our needs.

My thoughts turned to the investor-owned utilities. They were always having seminars and training sessions of one sort or another. In my travels around the country I proceeded to ask a lot of questions about the training programs conducted by each utility I was visiting. The more I dug into it, the more discouraged I became. They were professional; had really refined the in-house training concept. I just couldn't seem to break into exactly how they did it.

One of our managers suggested I contact the training director of our local utility. I met with him at their training facility. I shall be forever grateful to him, as he told us how to do it. He confirmed developing training programs was difficult at best. He suggested we start with developing retraining programs. As soon as he said it, bells and whistles went off! Hurrah, the solution!

I sat in on an equipment session and was thinking about the retraining approach, when I looked over and saw good old Virgil from Toledo. His equipment cost and equipment utilization were about the best in the company. He could teach the class rather than sit in it! I now knew how to start the retraining program. The basis for the program would be performance evaluations of each supervisor and manager.

We immediately initiated evaluation programs from every level of management. Account managers evaluated supervisors; division managers evaluated account managers. Each group of evaluators went over the evaluation with the person being evaluated. Everyone signed the form. I was awash with signatures! Each evaluator rearranged his schedule so they could spend time working with individuals, helping them improve their skills.

In the meantime, we redesigned our week-long seminars which encompassed every subject into several one-day retraining sessions, each encompassing just two or three subjects. People who were struggling in certain subject areas were brought in just for these specific sessions.

You can rest assured that good old Virgil was one of the instructors in the session on equipment. Peers telling peers how they do things and obtain results is extremely effective, and even more so when combined with people presenting the theory end.

Looking back, the solution to the problem seems so simple. We had been too close to the problem. The outside perspective from the utility training director showed us the right actions to take to arrive at the answer we wanted.

If your business is selling and managing people, and you spend your time organizing your efforts so you manage people effectively, you will be absolutely amazed at how much time you will eventually have available again.

After a while, you will find that fewer and fewer situations arise which cause you to react. A trained professional staff, which is an extension of you and your ideals, causes fewer problems.

Evaluating and training add substance and stature to your firm. It sets you apart from the others. It adds to your corporate culture. It certainly beats handling someone a set of car keys and telling them they have been promoted and to go get 'em!

Isolate strengths and weaknesses, then retrain. You'll be surprised how far the organization can travel.

Being a good manager means knowing when to let go and when to delegate. If you are ever going to grow your business, delegation is a must. Determine what each person's strong and weak points are. Then leave him/her alone in areas they are proficient and provide a backstop in the areas they are weak. The weak areas are the training grounds.

Did our training and organization program work? In six years, sales went from 52 million to 129 million. We never could have expanded that fast and as profitably if it hadn't.

Compensation

I am not going to go into detail on various compensation plans. There are as many plans as there are people. Books have been written on the subject and consultants abound. There are a few principles I have seen work over the years. Actually what works is common sense and KISS - Keep It Simple, Stupid!

No compensation plan yet devised is a substitute for management. Many plans have been developed consciously or unconsciously in the hope the compensation package will solve more problems than just remuneration. It won't work. Plans and policy are not substitutes for hands-on management. You can delegate, but you can never abdicate management duties.

Management should not expect results from a compensation package other than what it is supposed to do; compensate people fairly for their efforts. A compensation package which ignores people's basic needs is going to cause ongoing friction between manager and employee. Come up with a plan which does not address basic needs, or places basic needs in jeopardy, and you will come up with strife, headaches and management distractions.

I just read an article where the author had linked an employee's family hospitalization cost to the employee's sales quota. He alluded to having some difficulties with his people and this concept. I hope he has plenty of aspirin. The employees would probably react as follows: If you put my family and me in jeopardy, and make me nervous about their welfare and comfort, then I sure will set up a storm for you. Come up with a compensation package which is fair and equitable, and then manage me. You will get your money's worth, or you will replace me. Just don't tell me my family's needs are in a constant financial limbo and are dependent upon quotas.

If ever there was a case of knowing the answer and working the problem, this is it. In the case of a salesperson, you know how much must be sold to generate "x" profit dollars to cover his cost to you. His cost is defined as a total compensation package which meets his basic needs. If he performs better than his base, reward him further.

In the case of hourly personnel, don't pay too much attention to your compensation. What is happening in your local universe? What are other local industries paying?

Talk with the local employment office. Find out what other industries are paying by job group. Then, it isn't difficult to break down the information and translate it into your apprentice learning schedules and the job descriptions.

Catch 22

I pay higher wages and I squeeze higher profits.

Production efficiency is rarely factored into what we are willing to pay someone or even what we are willing to charge for someone.

Remember what I said earlier about White Nurseries charging twice as much as the competition, and always having more work than the competition?

The staff were paid according to their abilities, not to what someone thought the market dictated. Billing rates are secondary to quality performance and efficiency. The customer is smart enough to know when he is getting a good deal and the results he wants. He is not interested in what you pay your people; it's what your people produce that counts.

Twenty-percent companies have something else in common: they pay key people even if the snow is eyeball high and no work can be done or billed. Key people can count on being paid regardless. There is a minimum weekly income set. This nonproductive time is factored into the billing rate. The customer wants good production from experienced people, and is willing to pay for it.

One last thought on the subject.

You should also be willing to pay for performance above and beyond the performance level covered by the base compensation plan. Formalize the system, then reward and announce it at the beginning of the period. Always let people know what the reward plans are and the current status towards the goal or plan.

We were thrilled when we had field managers making more than the president and CEO because of exceptional performance. They made it; so did the company. Another golden rule perhaps?

Tell Them You Told Them

In public speaking they say: "Know your audience. Tell them what you are going to tell them. Then tell them. Then tell them what you told them."

So………

I have never met you, but I believe I know you.

The very fact you are reading this book tells me you are a twenty-percenter. You are interested in exploring every avenue to insure success.

I like to think I have been in sales and staffing situations similar to what you are in.

When I started out, I really did not know how to solve the problems of growing and staffing. At first I didn't even realize the problems in these areas which I would encounter. Then I became frustrated. I just wanted to design beautiful landscapes and then construct them. Why couldn't the market see just what a great designer and fine fellow I really was? Why didn't they flock in droves to enlist my services? Those other companies which seemed to have been around for years, how did they do it?

When I first found out I really wasn't in the design and construction business, but the customer service and people management business, I was resentful. I had always wanted to be around trees and plants. I couldn't think of one other thing I'd rather do. Finding out I had to spend time servicing the client to the infinite was not too appealing.

However, when it did finally dawn on me this was the only way to go, it became easier. There are no short cuts in developing sales by servicing the client. It was not very glamorous either, but it absolutely was the least expensive and the best way to build a business. That certainly captured my attention.

The aspect of having very small costs involved in obtaining new sales through service was something I could get excited about. After all, how much do a clipboard, some 3" X 5" cards and 500 business cards cost? I could take on the mightiest and the most venerated and succeed.

When I was running my own territory, I could maintain additional sales by utilizing the client's tendencies for one-up-manship. The clients and I could sell as much in one year as other territories could with two salespeople. The third year I was in the territory, I won third place in the company's annual sales contest, the fourth year I won second place, and the fifth year first place. Love those clients!

Whatever part of the green industry you are in, keep the services you offer the market simple. Don't try to be all things to everyone. The services you do offer should be things that you and your staff can effectively accomplish to the client's satisfaction. Then just keep going back.

If you are not on a first-name basis with a client, you are not there yet. I had one client in Watkins Glen, New York tell his wife in front of me to invite me to their cocktail party the following Saturday. He said, "You had better invite him, he is always here anyway, running around with those hand pruners of his."

People want attention. They respond to it.

Your staff is the same as your clients. They want and require just as much attention as the client. They will respond in the same enthusiastic way the clients do, if you show genuine interest.

Once you get over the shock of finding out you are really in the business of selling and managing people, and once you learn to successfully deal with employees, you will find things begin to become fun again. You might even find the time to do what you thought you were going to do when you started the business!

Example #1: Fill in Letters

Letters are preprinted using the office computer or word processor. A blank space is left in the middle of the letter so the office staff can insert your "personal" message and salutation.

Dear Mr. and Mrs. Virgil:

People often ask if it is really necessary to feed trees and shrubs. They will often point out trees in the forest seem to do very well, so why feed their trees?

In the forest, leaves accumulate on the ground and decay. This process adds nutrients back into the soil where the root system can absorb them. There is a complete cycle of leaves falling, and decay producing nutrients which the tree utilizes to produce more leaves.

Man interrupts this natural cycle when we rake up all the leaves every fall. We further compound the problem by maintaining lawns that compete with the trees for food. Feeding the lawn doesn't help the tree. Lawn feeding does just what it says; it feeds the lawn. Add in pollution and you have a serious problem with trees trying to grow outside their natural forest environment.

(THE ADDED MESSAGE)

("We suggest you consider a feeding program for the Sycamore Maple by the patio and the Oaks and Dogwoods along the north and rear property line.")

Money spent on feeding does more for all things growing than any other service we offer. I will telephone you in a few days to discuss the details of our feeding program.

Sincerely,

Harry Twenty-Percenter

Example #2: Door Hanger

Foreman: _____ Date: _____

Arrived: _____

Left: _____
(does not include dumping)

We Did: _____

_____

_____

_____

We Noticed: _____

_____

_____

Dear Mr. Twenty-Percenter:

- ❏ Crew did a great job.
- ❏ Stop by and have a cup of coffee any time.
- ❏ See me about what the foreman noticed.
- ❏ See us next year.
- ❏ I am mad!

COMMENTS: _____

NAME: _____

True Life Stories From The Real World

✦ *I Bought Out Part Of A Company*
Where I Worked

Rick is a contractor extraordinaire. It's in his blood. Almost everything which can be experienced in this business has been a part of his history. He has paid his dues and then some. But, unlike many who have had similar experiences, his company is still alive and well.

He started when he was just 17 years old. He cut his construction eyeteeth running hydro-mulchers, and tractors. For two years he went to the Colorado School of Mines, a highly respected engineering school. Though he had the intelligence, he didn't have the patience to be an engineer. He likes to build.

His hard-driving personality and ability took him a long way from the hot seat of tractors and the dust of construction jobs. At the age of 27, he became a vice-president of operations for one of the largest landscape companies in the country. His responsibilities included the coordination of all men, equipment, and materials to get all the work accomplished.

But differences in opinion soon surfaced between Rick and another key employee. After several attempts to solve the differences, Rick felt he needed to put pressure on the situation. He put forth an ultimatum and was fired.

Rick was offered some existing contracts and some equipment in exchange for his stock. Thus he became one the nation's largest landscape and site development contractors in the country.

Little did Rick know within three short years he would buy out the rest of the company which had let him depart. It would not be the first or the last company he would swallow up on his own. One year after he purchased the first company, he took on another, this one, he thought, by necessity. He loaned another nursery/contractor friend in town $50,000. Thirty days later, the friend needed more money to stay afloat and to pay back the first $50,000. Rick sent his accountant in to look at the books, and soon discovered the company he had loaned the money to was broke, and just did not know it yet.

In order to preserve his initial investment, he took the company over based on their inventory, equipment and people. However, the inventory was not even worth fifty cents on the dollar of its valuation and he lost almost $700,000 on the deal.

I asked Rick what he thought about those buyouts and whether he would do it again. His response was an emphatic "NO"! and he stated, "You usually buy out a company for two reasons: the people they have, and the work they have on their books. My experience is often the work you get is not very good work and most of the people leave in a short while after you purchase the company. In fact, only three or four people are with me from those buyouts. I think this is possibly for three reasons. First, the company philosophies are totally different and they do not care about our company philosophy. If they did, they would have come to work for us in the first place. Second, many times when a company is going down, these people's attitudes can be soured and they just want to put everything they had to do with the experience behind them and get a fresh start. Third, they come into a bigger company and fewer chances for advancement, since many key people are already in place."

But in the midst of all of this, another event occurred which would dramatically affect Rick and his company. Another friend had ventured into strip coal mining in Oklahoma. Thousands of cubic yards of dirt would be excavated to get down to an 18" or thicker layer of high grade coal that was to be shipped around the world. This friend invited Rick to come down and mine some of the mine leases that he had. Rick checked it all out, and it looked good until he got there with new equipment which had a $30,000 payment due on it every month. Then he discovered the lease was bad and there were no buyers lined up to purchase the coal once it was mined. In order to survive, he lined up his own leases, found his own buyers, and began a six-year coal mining venture.

According to Rick, the operation was profitable and successful until 1980. In 1980, the energy crisis had raised coal prices to a level which made the mine a success began to subside. For two months the mine was shut down while Rick worked on a way to stem the $150,000 per month loss which was now showing up on the books. He sold off $3,000,000 worth of equipment to reduce his debt load and equipment payments. He then started it back up with reduced labor and only $1,000,000 worth of better equipment. But the company was still marginal at best, and it was beginning to drain resources from the rest of the company.

As a friend, I was familiar with Rick's company during those days. One of the difficulties I feel existed in those days was the construction side of the company began to be very negative about the mining side of the operation. Losses were occurring in both sides of the company, possibly due to Rick being so involved in the coal mine operation he did not have time to properly supervise what was going on in construction so it was losing money. Regardless of the reason, when the coal mining operation was no longer their venture, I saw a dramatic shift in attitudes of certain key people who helped in the recovery of the company to its present condition.

However, disposing of the coal mine would be another very costly experience. Although it was currently losing money with no immediate change probably because of coal prices, the mine had made money in the past and thus its assets and net worth were of considerable consequence. The manager of the mine at the time found a "group of investors" who would buy the operation outright. Since it was a separate corporation from Rick's construction company a deal was worked out. However, the investors paid very little of the $1,100,000 net due him. They took it into Chapter 11 bankruptcy and then began to use its assets to pay their salaries and work under the table deals while the mine continued to operate under the Chapter 11 protection. Of that original note, Rick has seen only $500,000 come to him, and he doubts he will receive much more. From his experience he has come to see how the bankruptcy laws of this country need to be changed.

Rick had built his company through acquisitions, and the ability to go out and get work. But as he looks back, there are things he is glad he did, and things he wishes he had not done. All of them may surprise you in the light of what he has been through and what he has accomplished.

There are three things he is glad he did. He is quick to say through many of his struggles he turned his spiritual life around, and discovered that his relationships to God and to his family are more important than his company. Contracting can become a dog-eat-dog business which can turn someone into a workaholic. Tempting opportunities to cheat and get rich quick are often within reach. Rick is truly as honest a contractor as can be found, and he credits a lot of his success to this fact.

He points out every company has a company culture and philosophy on how it will do business. That philosophy comes from the owner. If your employees see you cheat or fit the facts to suit you, they will do the same, plus a little more. That kind of unwritten policy and philosophy can become very debilitating to a company's reputation and ability to be successful.

The next thing he is glad he did was to focus on being a people company. The truth of this is in the fact he has people who began with him and are still with him after fifteen years. He also has one of the best drug and alcohol abuse programs I have ever seen. The program does not just police people and find the abusers, it also attempts to rehabilitate them. He also has an Employee Stock Program in which he is attempting to sell the company to the people that have helped him build it, his employees.

Rick is quick to praise his employees for a job well done or positively confront them for a job which has not been well done. He has also pioneered some of the people ideas found in the section "Food for Thought" (page 549) having to do with recognizing employees for a job well done, as well as employees newsletters, to keep them informed of what is going on in the company.

Another thing he is glad he did was to divisionalize his company. His ten- to-fifteen-million dollar a year company has eight different divisions, each one with a manager responsible for making his division (or profit center) profitable. This allows Rick to focus in on those managers while they focus in on doing the work. It also gives him the feeling that the whole company is not resting on his shoulders.

But there are certainly things he wishes he had never done. We have already mentioned one but there are several which are just as vital. The most surprising is if he had it to do all over again, he would not want to have single ownership. In fact, when he started the company in 1975, he took in a corporate partner who had the option to have 50% of the company. In 1978, the partner exercised the option and then left. Rick wishes he had done a better job of selecting someone who would have stayed with it for the long haul. He has replaced this loss, and fulfilled his desire not to manage it alone, by having a very active board of directors and a group of division managers.

He also wished he had realized a company's ability to perform work has nothing to do with its bonding capacity. Several years ago, he took on a four-million-dollar-plus project. His bonding capacity was not used up, so he made the mistake of taking on another job of the same size. The combination of these two large projects at the same time proved difficult.

He also feels he has made the terrible mistake of hanging on to people too long. There are people he has had to let go, not because they were not good people, but because they did not fit what the company was doing at that time. He used to make the mistake of "finding" a place for these people until he realized he was doing both of them an injustice by that kind of action.

Rick has also experienced the upside and the downside of "burnout." He is a very intense person and sometimes the extreme hardships of this business have taken their toll on him emotionally. However, Rick has been quick to sense this and has found several remedies for his "burnout." Often, the remedy is to personally take on the supervision of a job, and get out close to his roots on the seat of some piece of equipment. Other times, it is to spend time with someone he looks up to as a mentor in the business, and talk through some of his frustrations. Whatever the cure, he has had to apply it more than once. Which proves, no matter how long you have been in it or how much you think you know about it, this business is always tough and exacting upon those who choose to navigate its waters.

✦ *I Grew Faster Than The Company Did*

Ben is the kind of guy every man would like to have as his son: tall, good looking and a tight end on his college football team. An Honor Roll graduate with degrees in Business and in Real Estate. Everyone who meets Ben likes him and wants to help him.

While in college, he met a friend who was a working fool. The friend attended all the classes, studied, and ran a substantial (for a college student) landscape maintenance business. Ben also had some maintenance business clients, but he was more into college. Even back then, Ben's talents would surface when he got a "big" job ($3000-$4000) that he felt was too much for him to handle alone. Consequently, he and his friend joint-ventured the job. Thus begin a relationship that exists to this day.

After graduation in 1983, Ben went out to apply his trade of real estate by serving an internship in the marketplace. His disarming charm and likability would begin to open doors for him and his future. A developer friend listened to Ben's dream of being a real estate mogul, and advised him to start a business that would raise cash for him to use to amass a real estate empire.

He teamed up with his old college friend and they started out into the marketplace. They quickly recognized that Ben could sell snow to an Eskimo and was a marketing genius. He could bring in the work, and his friend was to get it done.

Within six months they discovered something else. The ability to bring in the work quickly exceeded the needed capital to do the work and keep the company afloat. Then they discovered something else: Ben had the ability to go out and raise capital just as efficiently as he raised and brought in the work. Everything looked great. The company had a marketing/money raising genius who was fueling the company with money and work.

However, getting work and the money to do it is only half the battle. The company and Ben's abilities outgrew the field operation's ability to get the work accomplished in a profitable manner. Ben brought it in at a head-spinning pace and the field couldn't keep up. Tension set in upon the company.

Ben felt like he was doing his part and more, and he couldn't understand why field operations weren't keeping up. In the meantime, field operations had the attitude of "all right, already, slow down the work."

By this time Ben had gained the majority of stock in the corporation due to his capital raising success. He felt if they just hired more people and put more money and more people into field operations, it could solve the problem. However, all it did was bring across their path some of the "best and the worst" in field operations. The new people began to cause even more problems for the company.

In 1988 the company would pay dearly for this very problem. They hired a project manager who assured them he had the ability to bid and manage a two plus million dollar public works type project. They believed him and proceeded to successfully win a bid for such a project. Unfortunately the person could do neither bid nor manage the things he claimed and the project became a major fiasco for the company.

By this time the company was in the red by seven figures. In most situations, this would spell the end of the company. But once again, Ben's ability to handle such situations rose to the surface. He worked with his investors who ended up taking the majority of the hit. He has continued to operate with a personal goal of paying them back, even though it is not legally required.

Today is a different day for Ben and his friend. They are doing more than of five million dollars in sales, in two cities. But they have learned some very important lessons.

Most importantly, you must keep the amount of work coming into the company in direct relationship to your field operation's ability to perform the work. I have often said contracting is a balancing act. The key is not how much work you get, but finding the right balance. To get more work than the field can perform is just as disastrous as not getting as much work as the field can perform.

Sales and production must be viewed as two horses pulling the wagon of your company. If they are to work as a team, they must keep together. If one gets ahead of the other, you will have trouble.

Another lesson to learn is being able to get money through bonding doesn't mean you should use it. HUD was crippled for a time due to its program of non-qualifying loans. The program said "common sense says you should not qualify to get this house, but we will let you have it anyway." Given that kind of unhealthy permission, people ventured out beyond their ability resulting in thousands of houses in foreclosure. It is the same with business. Every company, independently of any bank or bonding company, must decide what is best for it and then, no matter what is offered, say "NO" if it is beyond their limits.

But Ben will tell you of some very positive things which have happened through all of this. He has found his niche, and his niche is working with and negotiating with owners/developers. He has finally built a field operation which can keep up with his sales and marketing ability. Along with that, he has found a better awareness of the balance which must occur between sales and field operations.

But most of all, Ben, is more careful about what he believes. His youthful naiveté is now replaced with a more careful evaluation of people and their promises.

Troubleshooting the Estimating and Bidding Process

The Importance of a Strategy

Business, in many respects, is like a war. You, as the owners and managers of a contracting company, are the Joint Chiefs of Staff, making plans for an approach to the battle called "getting and doing" contracting jobs. Over the years, as I have worked with contractors, I have discovered the extreme importance of making these plans; of having an "estimating strategy."

A good strategy does four things for your company. First, it enables you to know how every dollar you spend is going to come back to you through the categories on the estimates and bids you produce. One of the most beneficial things about having an estimating strategy is costs which do not occur on every job, or those which are hard to pinpoint, will not fall through the cracks between what you pay out and what people pay to your company. A well-thought-through strategy will allow you to track each item in your checkbook carefully and completely, and to recover costs in the best possible ways.

Should you recover a certain cost through job costs, labor burden or overhead? The decision should not be based on which is easiest for you—though this is a consideration—but according to where the cost can be recovered in a proper way. Sometimes, mathematically, it will not work to put certain costs in certain areas. I have found contractors losing tens of thousands of dollars per year because they have put things into a wrong category, or have combined certain things which should not be combined.

One Example: Estimating the costs of equipment has always been a great problem in contracting. If these costs are considered part of overhead, jobs which use very little equipment are penalized and have to pay—through overhead—more than their "fair share" of the equipment costs of the company. If you estimate your equipment costs on a job-by-job basis, you must still put any equipment used for overhead purposes (e.g., the owner's car or truck, the field manager's car or truck, etc.) into your overhead figures. If you don't figure your equipment costs this way, you will have money falling through the cracks.

Second, a good estimating strategy will compensate for the variables which exist from job to job. You must recognize that no two jobs are exactly alike. Every job has different site conditions, requires a unique approach, and entails different costs, regardless of whether or not the materials are the same. A good strategy will give you the flexibility to compensate for the variables. I have worked with contractors who have made money on one job at certain unit prices, but when they used those same moneymaking prices on another job, they lost money. They knew things were a little different on the second job, but did not know how to compensate properly for those differences. In future chapters I will go over a system of estimating which will give you that ability.

Third, a good estimating strategy will give you the ability to control your jobs after you get a contract. That is, you will be able to make what happens in the field prove your estimate. This business is no less than a two-punch business. Punch number one is putting out a good competitive bid. Punch number two is getting the work done for the amount of money you estimated in your bid. A good estimating strategy will give you the ability, to take the estimate, to make it become reality in the field. The next section of this book will talk about some of the principles which can make that happen. It really all begins with an estimate done in such a way it can be used to control the job in the field quickly and accurately.

Fourth and finally, a good estimating strategy will give you the ability to make sound business/financial decisions. That's right! Your estimating strategy is one of the most important places to gather information for such decisions. We are a very goal-oriented, forward-planning consulting firm. We believe in setting up an overhead budget for a year in advance of the current year. We then set up overhead recovery percentages to recover those costs based on how much work we expect to contract and complete with our work force and the budgeted overhead. Those kinds of projections and plans give us a plumb line of where we are going. But, things never work out exactly as we plan. Therefore, a good estimating strategy helps us determine what changes we will need to make in overhead costs, labor burden benefits, or salaries so our company will remain profitable and competitive in our marketplace.

Just as in battle, officers and infantrymen observe signs telling them to change tactics, you will see the effects of owners' budgets, other contractors' bids, and shifts in the economy that tell you to adjust your strategy in order to be successful. You will not be able to establish one estimating strategy and expect it to work for you year after year. In fact, the most successful companies rethink their strategies annually and monitor them after several bids.

Still, an estimating strategy will help the manager or owner make sound financial decisions. Too many management decisions are based on emotions or on short-term situations. For instance, a manager may want to hire additional people, or pay more benefits to his people, or give them raises because his employees come in and ask for these things. The manager, based on feelings and without proper ability to explain why these things can or cannot be given, grants or denies the requests. Other times, a manager or owner may go out and buy a computer or other pieces of equipment without basing the purchase on any sound decision-making process. A good estimating strategy will give you such a system, such a process.

For example: Let's say you want to hire another secretary. Everyone *seems* busy and it *seems* you could use her services. But, an apparent need should not be the major reason to hire her. Instead, you should base your decision on what you know from your estimating strategy. You established an overhead budget that did not include another secretary, and you are acquiring jobs based on the budget. Will hiring her allow you to get enough additional work done to pay her salary? Or, can you raise your prices enough to pay for another secretary by doing the same amount of work with an increased overhead charge, and will you still get your projected amount of work for the year? Your estimating strategy shows you that you cannot afford another secretary and meet your goals. So, can some of the things she would do be done by existing staff or even not be done at all? An estimating strategy will help you make sound decisions based on numbers, not emotions.

I close this section with a diagram I use around the country to demonstrate this point. On the next page you will see two triangles. Triangle A is inverted and shows how some contractors go out and get jobs with whatever price has worked before. They then work to keep costs as low as possible, keep overhead as low as possible, and hope at the bottom there will be some profit left. The problem is they have no idea just how low costs and overhead must be in order to make a profit. And, if things change in pricing, or in their company, they have no idea what effects the changes will have on the bottom line.

Triangle B shows someone with an estimating strategy. The same triangle is turned over. This person starts with estimated costs, budgeted overhead, and a reasonable profit. They *know* what they must do in order to make a profit. When the smoke clears after the battle, they will be on top.

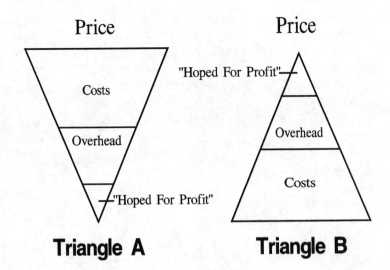

Cost/Profit Estimating Systems: Triangle A is based on old prices and costs and Triangle B is based on an estimating strategy.

An Overview of the System

Here is an overview of an ideal estimating system for any contractor who wants to approach estimating with a strategy. At the end of this section (page 434) is a diagram illustrating all the items discussed below. Detailed information on each aspect of the system is presented in the sections which follow.

When I estimate a job, I do it in three parts. The first two involve estimating: arriving logically at the costs of any job to your company. The third part develops these numbers into a bid, which is the final dollar amount for which I can do a job, and the price I submit to be considered for the job.

Part One

Produce the Product: That's what I call the costing of material, labor, equipment, and subcontractors to build or produce "the product"—the job the owner or manager wants done. Costs of material will be actual costs, not based on any retail or "before discount" prices. The cost of labor is based on the number of hours it will take a known crew to do the job, multiplied by their actual wages, plus any overtime compensation, and a "fudge factor" built in to cover the unexpected or time wasted. The cost of equipment is based on the number of hours your equipment will be on the job, multiplied by its costs. Subcontractors will cost out at the actual price they give you for their portion of the work.

Part Two

General Conditions can be defined as job overhead and includes things that are not part of producing the actual product, yet neither are they a part of the general administrative overhead. I estimate these costs as part of each particular job's costs under

General Conditions, which—like Part One—break down into material, labor, equipment, and subcontractor categories.

Part Three

Markups are the part of the system that really requires a contractor to think strategically. It plays the greatest role in a contractor's estimating. Here, I begin by adding any necessary taxes to material, and labor burden to labor dollars. Labor burden is labor overhead, including all payroll taxes, insurance for labor, and benefits paid the labor force. This area must be monitored very carefully, and a contractor should keep in mind what other contractors are paying in benefits to their people.

The next area I work on under markups is overhead recovery: how are the dollars you spend on overhead going to come back to your company? Again, this is a *very* important area of the estimating strategy. Many times, if a contractor is not successful at the bid table or is not making money, it is because he is not re-working—or understanding—this part of his strategy. For instance, a contractor may have more overhead budgeted than he needs to do the work he intends to do. Or, he can do more work with the overhead he has budgeted than he thought he could. Or, he may need more overhead to handle the work he is contracting. In these cases, he will need to adjust his budget for overhead recovery.

First, I establish the yearly overhead budget and project an amount of work for the year (see pages 461 and 465 for the methods of doing this). Then, I work up an overhead recovery percentage for each of the areas of cost - material, labor, equipment, and subcontractors - to get my overhead recovery figures.

Now those overhead figures must be tested in the marketplace. Are your bids high or low? Are you getting jobs? It is crucial you adjust your strategy, *and* those overhead figures, as soon as you see how you are doing with your bids. Also, take a look at your actual overhead versus your overhead budget. For example: You will need to adjust your strategy if you see you are able to do more work within your current overhead budget than you thought you could.

The next item to be considered in the bid under markups is profit. Again, this is another important area of strategizing (see the section on page 483).

The final item to consider is the "Ding-Dong Factor." This factor is used when you need to add money to compensate for a bad architect, owner, developer, or situation. As you can see, this system lends itself well, not only to developing a strategy, but to being able to change your strategy if necessary, at any time and on any job.

The following sections will help you focus in on the details of materials, equipment, labor burden, subcontractors, and overhead.

✎ *An Outline of the System* ✎

I. Produce the Product

Material	**Labor**	**Equipment**	**Subcontractors**
Actual Cost	Production	Production	Actual Cost
	Hours	Hours	
	Average wage	Cost	

II. General Conditions

Material	**Labor**	**Equipment**	**Subcontractors**

III. Mark-Ups

Tax on Material   *Labor Burden on Labor*

Overhead Recovery

Material	**Labor**	**Equipment**	**Subcontractors**
Percent	Percent	Percent	Percent

Profit

"Ding-Dong Factor"

An outline of the ideal estimating system.

Materials

When you are dealing with materials for a bid, the goal is to come up with the amount you will write checks for in the end. There are three things you must consider.

Do not list what you take off the drawings. Rather, list what you are going to buy in materials. I love computerized estimating. However, I think it can make estimators into lazy brains. This is really true when it comes to the new digitizers. Digitizers can take the material off of a drawing and put the quantity into a computer-estimating program without the estimator even having to insert the quantity into the bid.

As an example, I have seen landscape estimators produce bids that show 1,113.5 feet of 2-inch pipe. This is what the digitizer took off the drawings down to .5 feet. I ask the estimator, "Where are you going to buy .5 feet of pipe? In fact, since pipe comes in 20-foot lengths, where are you going to buy 13 feet of pipe? And, if you send out 1,120 feet of pipe to the job, you are going to get a call before that job is finished. Do you know what that call is going to be about? They need more pipe on the job. They will need more pipe on the job because two things which need to be on the same pipeline were 17 feet apart. The field person went and got a 20-foot-length of pipe, then cut 3 feet off the end to get their 17 feet. What did they do with the leftover 3 feet? Put it in their back pocket to see if they could use it somewhere else on the job? No, they threw it into the trench with the other 17 feet and covered it all up. Who paid for the 3 feet of pipe? You, the contractor paid for that 3 feet." That is why you never put into a bid the exact quantity which is taken off the drawings by a digitizer or by hand. Rather, put down what you will buy.

If someone is doing a good job of their take offs in this area, they should be able to order the material right off the estimate sheet and never get a call from the field saying they need anything. If such a call does come in, the one who did the estimate can know immediately they did not take off the material properly.

Third, in estimating materials, consider any guarantees or warranties you must provide to the owner. When you are asked to guarantee something, especially plant material, you are going into another business: insurance. You are insuring the livability of plant material in the environment in which you plant it.

If you are going to be in the insurance business, you'd better think like an insurance person. Let's say that I, a forty-something male with no accidents or tickets for the last ten years, was talking to my insurance man about insuring me to drive a 1998 truck. While we are talking, a 16-year-old male with pink, yellow, and red hair drives up in a brand new Corvette. The ink is still wet on his driver's license. He wants to know what insurance will cost him. Will the insurance person quote us the same price for insurance? Of course not. Why? Because the 16-year-old male is a much greater risk in that Corvette than I am in a 1998 truck. You have to make similar distinctions when guaranteeing materials.

I suggest that you determine the risk or loss factor on each kind of material that you will install, then add a percentage for loss to the wholesale price of that material before you put it on the estimate sheet. For example: in landscape contracting, where risk of loss is high, a shrub that costs $7, but has only a 5-percent risk factor would be priced on the estimate sheet at $7.35. A tree that costs $80 but has a risk factor of 30 percent would be priced on the estimate sheet at $104.

Remember, the goal of handling material on an estimate is to come up with the amounts you will finally write checks for, including guarantees and warranties.

Labor

I have told over 70,000 contractors across North America I have yet to meet the contractor I can't make profitable if he or she will focus on just two things: the control of labor and overhead. Sad to say, most contractors focus on something else: getting work and growing.

A major bonding company—an insurance company which guarantees that a contract will be performed at the original bid price even if the contractor goes bankrupt —did a survey of contractors who had declared bankruptcy to see why they went bankrupt. Do you know how many contractors they found who had gone broke due to lack of work? NONE! Do you know how many contractors had gone bankrupt because they had taken on too much work too fast and too cheap? LOTS OF THEM! A contractor's primary focus must be on the control of labor and overhead. I will discuss the control of overhead in a later section. Let's go into the control of labor here.

The labor portion of any estimate is the greatest risk in estimating. I have never stayed up nights worrying if I had enough for materials in a bid. If you can't take off material from a drawing, get out of town. I have never stayed up nights worrying about subcontractors. If I know what the subcontractors have included, and if I have covered what they have excluded, I can sign them up and get them to do their work. Equipment follows labor, so if labor is right, then my equipment will be right. However, I have stayed awake many a night worrying if I had enough or too much labor in an estimate.

When I deal with labor on an estimate, I never deal in dollars and cents. Rather, I use production hours. Production hours are the most stable way of estimating labor for the long term.

If I pour 30 square feet of concrete per hour and I have 300 square feet to pour, then I need 10 production hours, or if I can put in 5 feet of water line per hour and I have 50 feet total, then it is 10 production hours.

I deal in production hours for three reasons. First, the dollars-and-cents labor costs become antiquated very quickly. You could show me labor for a job you did yesterday, but as soon as you give anyone on that crew a raise, your costs will be antiquated. If you had a foreman with five low-priced laborers—who lower your labor costs per hour—doing the work on one job, but on the next job that same foreman has only two low-priced laborers, your costs will be antiquated.

I know how many production hours are necessary to form and pour and finish small concrete slabs. I have been using those same production rates for years. Workers form and pour and finish small concrete slabs the same way today as they did 20 years ago. Do you think labor costs are the same today as they were 20 years ago? Of course not! What has changed? The cost of labor per hour. I can use the same production rates and be current on my labor costs by multiplying the production hours by today's average wage.

The second reason I use production hours is it is easier to vary production hours than to vary dollars-and-cents labor costs. For example, I can plant a tree every hour and thirty minutes if I can get a truck with the tree on it near the hole. I have ten trees to plant, five of which I can get near with a truck. I will need to tractor five others into another area and carry them the last 50 feet. It will take an extra 30 minutes on those five trees. I can vary my final production hours by figuring 15 hours to plant in normal conditions and another 2 1/2 hours for the five that are not normal. Total: 17 1/2 production hours.

The final reason that I use production hours is field people think production. Let's say you have done a bid based on dollars-and-cents labor costs and you have arrived at a total labor cost of $900 to install a furnace. If you go to your foreman and tell him to get the furnace installed for $900, what will he say? "What do you mean by $900? Who is paid what?"

But, if you have arrived at your labor costs by production hours multiplied by the average wage and you have calculated 80 production hours, then you can tell him with a four-man crew working ten hours a day, he has two days to do the job. Does your foreman understand this kind of goal? Of course he does, because this the way field people think: How many people are you giving me, and when do I have to be done?

I love job costing, but the problem with job costing is when you finally get the figures, the job is over. I have clients who know how much material needs to be installed in a day to stay on schedule. Based on the production hours in the estimate to install material, they can determine how large a crew to send out to the job to finish the work in a day. They can check on a job in the middle of the day and quickly see what has been installed and the hours used to see if the job is on target. If the crew does not get all the material installed in a day, it is obvious to everyone they are overrunning the estimate. However, this kind of labor control only happens when a contractor deals with labor on his estimate in production hours, not in dollars-and-cents labor costs.

Once I have arrived at the production hours for a certain type of work, I multiply those hours by the average wage for this crew. To find the average wage for bid purposes, I put together an average crew with their different hourly wages. These wages totaled will tell you the cost of this crew for one hour's work. Then you divide this amount by the number of people on the crew to arrive at the average wage.

You then add a factor for overtime to the hourly cost. For example, if you are working your crews 50 hours per week, you are paying them time and a half or for 55 hours per week. This means you are paying for 5 hours of time each week from which you are receiving no production. In order to compensate, you will need to add 10 percent to the crew's average wage. (Five hours is 10 percent of 50 hours).

To this figure I also add a "fudge factor." This number compensates for the difference between the how long you think something will take and how long it really takes.

Say your foreman comes in and says he can't find the key to the skid loader. The crew looks in the ash trays of the trucks and their pockets. After fifteen minutes with four people looking the crew finds the key. It was in the ignition of the skid loader. The crew tries to start the skid loader, but it won't turn over. They ran the battery dead. Someone is sent to get the jumper cables out of a truck, but someone else has taken the jumper cables. The foreman sends someone to the nearest discount store to get jumper cables. (Sound like one of your jobs?) The crew hooks up the jumper cables, but the skid loader still will not start. The gas tank is empty. Someone goes to the company truck to get their extra 5-gallon can of gas, but someone else forgot to fill it up after they used it, so someone else heads off to a gas station. Four people just spent an hour to get the skid loader started. Where is that in your bid?

Or, here is my favorite. Rain clouds move over your job and it begins to sprinkle. However, there is blue sky all around. Your foreman looks up and sees all the blue sky and determines the storm will not last long. Then he says the worst thing he can say, "Let's get in the trucks and wait it out." He was right, the storm only lasted half an hour and they were going to get out and go back to work, but their favorite song came on the radio. It's another ten minutes before they get out of the truck and go back to work. Where is that time in your bid?

I suggest if you know your crew's production rates and are very comfortable with them, you still add a 5-percent fudge factor to the average wage. If you feel moderately comfortable with them, add 10 percent. If you have never estimated in production hours before—and you will be trying to determine them as you estimate—add 15 percent until you get more comfortable with the process.

Now, it is very important to add this fudge factor to the wage and not the hours. The reason is because I want to hide the fudge factor from two different groups of people. The first is the owner or manager. Let's say you have a bid with 1000 production hours and you would add 10 percent or 100 hours for the fudge factor. As you begin to think about the job, you might say "I want this job. I need this job." With these thoughts in mind, what might you take out of that bid? The 100 hours of fudge factor. You get the job. How long does it take to do the job? 1,100 hours. Crews are still going to misplace the key to the skid loader, have trouble starting it, and run to get jumper cables and gas. It is still going to rain and workers will still sit in the trucks. I know contractors, if they can see those fudge factor hours, they are going to mess with them.

The other group of people that I want to hide the fudge factor from is field people. Can you imagine telling a foreman that he has 1000 hours to do a job, but you have also added 100 hours for wasted time and mistakes. They will say, "No problemo, I'll take care of those 100 hours."

I do not add the fudge factor to the hours but to the average wage so that neither the owner or manager nor the field people can see it and mess with it.

The following is an example of the average wage calculation:

Foreman	$12.00/hr.
Leadman	10.00/hr.
Laborer	8.00/hr.
Laborer	6.00/hr.
Total Crew Cost	$36.00/hr. ÷ 4 = $9.00/hr. (Average Wage)

Average Wage	$ 9.00
10 % Overtime	+ 0.90
Subtotal	$ 9.90
10 % Fudge Factor	+ 1.00
Average Wage	$10.90/hr.

Equipment

Equipment is something which should never be estimated on a percentage basis. Some contractors take a percentage of labor and come up with an equipment cost. They have arrived at a magic number by figuring if in a normal year their equipment costs are $60,000 and their labor is at $200,000, then equipment is 30 percent of labor. Sounds reasonable, doesn't it? But let's set up a scenario and see how it works. Let's say you are bidding two jobs today. One is labor intense and will only require your light, inexpensive equipment, and the other is equipment intensive and will require your more expensive equipment: backhoe, trenchers, tractors, etc. The labor on both jobs is $80,000. The actual equipment on the first is $10,000; on the second it is $30,000. If you were using that 30 percent, you would bid the first job's equipment at $24,000, too high a figure which may keep you from getting the job. On the second job you would cost equipment again at $24,000, which would be $6,000 too low, buying you a job you will wish you didn't have.

The most common way contractors bid equipment by a percentage is to include it in their overhead. The same example I used above will also hold true in this case, because you will just be recovering equipment randomly by percentage. Estimate equipment and cost it like labor. It is far too big a variable to be using a percentage for its cost recovery.

Here are a couple of ways you can come up with figures to use for your equipment. The first is an accountant's method using depreciation, maintenance, and gas costs. This is, by far, the best method to use. The illustration on page 446 is a form you can use to do this work on your own equipment.

Let's use a pickup truck as an example. Let's say you pay $9000 for a pickup. You are financing it over three years and the total cost to you, including purchase price, tax and three years of interest is $12,480. You have established a lifetime expectancy for your pickup of three years. At 40 hours per week, that's 2,080 hours per year or 6,240 lifetime hours. If you divide the cost of $12,480 by the 6,240 hours, it equals $2 per hour you will need to charge to replace that pickup.

Next, we want to figure its maintenance, license, and insurance costs. Because this varies depending on your location and how you approach *your* maintenance, I suggest you estimate three years of cost. Don't be ridiculously low. Put the pickup in the hands of an employee and be as honest as you can about how they will treat it and how much it will cost you to keep it on the road. Let's say you have arrived at a three-year cost of $6,240. Divide that figure by its lifetime hours of 6,240, which equals $1 per hour, which is what you need to charge to maintain the pickup.

Next, we want to figure its average gasoline consumption over a week. If you feel your pickup will consume 30 gallons/week at $1.20 per gallon, this equals $36 per week. At 40 hours this is $.90 per hour. The total cost is $3.90 for the pickup. Here, I always apply my favorite principle, round up. I always round my equipment costs up to the nearest .50 cents per hour. Therefore, I would use $4 an hour for the pickup. If you would rather use a per day price, simply multiply the hourly price by 8 hours to arrive at $32 per day.

I always recommend a contractor not run out and use those figures until he has put them to a two-step test. First call your local rental equipment company. See what rates they are charging for the same type of equipment. Be sure to see if they have included fuel. If they haven't, add your fuel costs to their prices. Remember, the rental people have included overhead and profit in their prices, whereas you are trying to arrive at costs *before* your overhead and profit. Consequently, your prices should be lower than theirs.

Realize these people are in the business of knowing the costs of equipment. They live and breathe life expectancies and maintenance costs. If your prices are considerably lower or higher than theirs, say by 30 percent, guess who's wrong? It's not them. Reevaluate your numbers.

Second, go to your local highway department and see if they have a book entitled *Equipment Rental Rates for Force Account Work*. Whenever a highway department wants extra work done on a job they do not ask you to work up a price to do it. They *tell* you to go do it and keep track of your costs.

Equipment Costing Calculations
...To Be Used on Time Cards

Item	Use	Purchase Price (Incl. Interest & Inflation)	+	Life Expectancy	=	Per Hour Cost (#1)	Anticipated Lifetime Maintenance & Insurance & License Costs	÷	Life Expectancy	=	Per Hour Maintenance Cost (#2)	Gas & Oil Per Hour (#3)	Total of (1, 2 & 3) Equals Per Hour Cost
Pick-Up	8 Hrs. Per Day	$12,480.00		6,240		2.00	$6,240.00		6,240		1.00	.90	3.90
Tractor	Off Of Hour Meter												

Equipment Costing Calculations Form to be used on time cards.

Subcontractors

Subcontractors are independent business people who will perform a part of your contract for a firm price. Subcontractors are *not* people who work under your direction for some hourly charge. Their employees are not under your direction. The Internal Revenue Service has very strict guidelines by which they determine who is a subcontractor and who is not. I would suggest you get a copy of those guidelines and study them.

Many contractors do not like to use subcontractors, because they feel subcontractors take a part of their business. I am a strong believer in expanding the talents of a company's people so the company can do more in-house. The more you can do with your own forces the more competitive you will be with your bids. It will also be easier for you to coordinate your jobs without having to depend on subcontractors and their schedules. But, I quickly recognize that there are things a company's personnel should not be asked to perform. Sometimes it is wise to take a job which has certain tasks you can do very profitably if you can subcontract the not so profitable portions.

How do you know what work to subcontract? Ask yourself two questions: Is this the kind of work my people are able to accomplish or will it stretch their abilities too far, and end up costing more for them to learn to do the work—and fix the mistakes they'll make while they're learning. Do my people have the time to do this work or will it interfere with other work or potential work which is more important?

I have had some contractors want to subcontract all or most of a job. They become construction managers only. In this business, I have yet to find anyone who does this with great success. This is still the kind of business where you will always have to do a majority of the work in-house with your own employees. Generally speaking, I have found that you must do at least 75 percent of the work with your own people in order to be competitive, day in and day out.

When you subcontract something, you must understand, ultimately, you will be responsible for the whole job, including the subcontractor's work. Consequently, you need to check out any subcontractors you intend to use to make sure they are capable, do good work, and will be around to cover any warranty work. Do not be afraid to ask for references and *call* their references. Be sure to talk with other contractors who have used the subcontractors to see if they did their work in a timely manner and cooperated with the contractor in order to produce a quality job.

Most contractors like to receive more than one bid from more than one subcontractor. If you cannot get two or more bids, then call some other people to get a "feel" for how much this kind of work should be priced. Try to meet with potential subcontractors you are using for the first time so you can get a feel for them as people.

One of the most important things you need to know when you are getting a subcontractor's price is what it includes and what it does not include. Make sure to get in writing a detailed description of exactly what they are performing on the job and what they expect you to do for them. For example, are they doing their own grading or do you have to do it? What kind of grade do they expect? Are they looking for you to move some of their materials with your equipment? If you must have performance bonds from your subcontractors, did they include this cost in their price? Any work you must do for them, or because they are not including it in their price, must be included in your estimate. You should receive a written proposal from them before you submit your bid. The price they give you is the price you put on the estimate sheet.

One other word of caution. Do not sign a subcontractor's proposal form or contract and use it as the contract between your company and the subcontractor. You should have a contract that you initiate and the subcontractor signs, which binds him legally to the contract you will have with the owner. If you sign a contract originated by a subcontractor you may run into problems because it may not include certain legal requirements contained in your original contract with the owner for whom you are doing the job.

Once you are awarded the job, do not delay in getting the subcontractor a contract, and getting his signature. He is obligated to perform the work at the agreed price, but you have not really covered yourself until he signs. Many a contractor has hesitated too long, and found that the subcontractor's costs have gone up or he has become too busy to handle your job. Then you will have to find someone else for more money.

General Conditions: Important Costs

Every contractor on every job has general condition items. I don't care how small a contractor you are, what kind of work you do, or how small your jobs are, you have general conditions—and you should be putting them on your estimates. If you are not, you may very well be losing money.

Some of you may be covering these items in your general or administrative overhead. If so, you are accounting for them by percentage. As I stressed in the previous section, they vary too much from job to job to be costed by percentage. You can distinguish general costs very easily, so you should count and cost them, reason them through, and estimate them on a job-by-job basis.

What are the general condition items that should be estimated on a job? Before we list some, let's first define what they are. General conditions are overhead-type items that are not applicable to the company as a whole. They can be applied to particular jobs in a reasonable fashion because they are only needed for one particular job.

I like to think of general conditions as "job overhead." General conditions are things which cost you money on a job but are not a part of the finished product the owner will have when you're done. You pay for them while you are doing the job, but they leave with you when you are done. For example, portable toilets are general condition items, which cost money on a monthly basis. You need them to do a job, but when the job is done, you take them with you. They are not a part of the finished product.

In a set of specifications it is very easy to see what general conditions are. Every set of specifications is broken down into three major divisions. There is the proposal or bid division in the front of the book. In the back of the book there is a large section that details technical specifications. This section tells you the details of how you must put this project together and the kinds of materials you will be required to install. Between those two sections are two other sections called general conditions and supplemental general conditions. In those sections are thousands of dollars worth of items that will cost you money on the job, and that you must provide to complete the project.

Here is a list of some of those items. I will key *some* of them with numbers and discuss them individually after I give the list.

1 ....Supervision
2 ....Mobilization
3 ....Cleanup (daily)
4 ....Toilets/Job Offices
        Trailers (Office & Storage)
        Dumpsters/Storage yards
        Temporary fencing
5 ....General job equipment
        Barricades/Flag persons
        Tests/Engineering
6 ....Gophers
        Pedestrian ramps
        Special insurance (i.e., builder's risk)
        Plane fare for out-of-town jobs
        Job signs/Safety requirements
        Hauling material to the job
        Temporary electricity
7 ....Temporary water
        Punch list items
        As-built drawing
        Photographs of construction
        Per diem to key employees
        Job crane
        Special job material

The list could go on and on, because it will vary on each job. The key thing is that you recognize the importance and the *costs* of these items on a job-to-job basis, regardless of whether you are a general contractor or a subcontractor.

I estimate my labor for each function based on actual production times. So, I need to pick up unproductive times such as times of job layout, organizing the workers, meetings, phone calls, paperwork time, etc., somewhere else. I do this in the supervision category of general conditions.

1. I figure supervision in two ways. If I am a general contractor, I put in the number of people (one, one and a half, two, etc.) who will be needed to supervise for the duration of the project. If I am a subcontractor, or a specialty contractor doing smaller jobs, I figure the number of days I will have people on the job. For instance, let's say I have 200 man hours on the job with a five-man crew working 8 hours a day. That's forty hours per day. Consequently, I will have people on the job for five days. I then estimate how many hours I will spend in supervisory tasks per day if I am running the job. If I feel I would be supervising for three hours per day for five days I would put fifteen hours of supervision in general conditions.

2. Mobilization is the time spent to haul your equipment to the job and to set up any trailers, fences, or storage yards. It will also include the reverse procedure. Again, I estimate the distance, traffic conditions, and/or unique difficulties in getting to the job. Then I put in the hours for the labor and equipment to perform this function.

   This item also comes into play if you are paying "glass time" or drive time for your workers to and from (or between) projects. If so, the number of people you pay, multiplied by the amount of time to get to the project from your office, multiplied by one or both ways, multiplied by the number of days you are going to the project, will give you your mobilization hours.

3. Every day, crews stop early to put away tools and clean up their work areas. Sometimes, there are not many tools and the work area is such it needs little cleaning. Other times, they are using a lot of tools and small materials or there are sidewalks, curbs and gutters and asphalt to sweep and clean *every* day. That's why I like to estimate this as a separate item.

   Again, the number of men cleaning up (usually all of them) multiplied by the amount of time to clean up (15 minutes, 1/2 hour, 3/4 hour, etc.) multiplied by the number of days you are on the job will give you your clean-up time.

4. I remember bidding a job once which was three miles long and ran on each side of a small town. When I bid that job, do you know what I saw? I saw a person working on the far end of the job having to go to the potty. I saw them get in a truck, drive through town, stopping at every red light before the toilet, which was at the designated yard and staging area on the other side of town. I saw them go potty, drive back through town, stopping at a 7-Eleven to get a Big Gulp so they could go potty again later. They then drove back to where they had been working. Then do you know what I saw when they got back there? Another person got in the same truck and went through the same procedure all over again. We could have had one truck busy solely as a "potty truck." So, when I saw this, I estimated five potties and put them up and down the

job in proximity to where crews would be working. IIt was unique to this job. But, because I was estimating item by item, I could pick it up.

5. I estimate my equipment in two places. General job equipment I put in general conditions. This is the kind of equipment which is going to be at the job every day the job is going on. Pickups are general job equipment. One-tons or two-tons can be general job equipment. Bobcats, tractors, generators, and fork lifts can be general job equipment, if you know they are going to be there for an extended period of time. I put them in for that time. If they do not fit in this way, I estimate them as specialized equipment with the function they are performing.

6. Gophers are people who "go for this or go for that." Some companies have a gopher in general overhead because they have someone who does this sort of thing throughout the company. It might even be the owner of a company. But, some jobs require this kind of person full- or part-time on a particular job. The job may be outside of the city in which you are located, or be of the size or nature which requires such a person just for this job. Again, I can estimate the person's time by the number of hours spent per day gophering, multiplied by the days we will be on the job.

7. In order to give you a feel for the magnitude of some of these items I want to share an experience of mine concerning this particular item. I remember estimating and getting a site-development job for a contractor who required us to maintain a large acreage of grass for thirty days. No sweat. I saw this requirement in the technical specifications and included in my estimate two men for thirty days to maintain the area. On the

thirtieth day, I met the owner's representative to sell him the job. After we shook hands, he gave me a bill from the local water company. It was addressed to our company—and it was for several thousand dollars. I asked, "What is this?" The owner's representative opened the specifications to a pre-marked page and showed me these highlighted words in the general conditions, "Contractor will pay for water during construction." This project was on a water meter. Thousands of dollars worth of water had been consumed to run the irrigation system day and night to bring up dozens of acres of grass.

Well, we fought it, wrote letters, screamed and hollered, but after months of no resolution, we finally paid the bill. Seven little words in the general conditions cost us thousands of dollars! "Contractor will pay for water during construction."

Look at the end of the section *A Sample Bid* and study some of the illustrations of bid sheets (begins page 495). These should serve to help clarify what we have just discussed and give you an overview of the process to this point.

Labor Burden

Labor burden can best be described as taxation on labor or on items which only apply to labor, and are not a part of the general administrative overhead. It has nothing to do with material, equipment, or subcontractors. Labor burden costs are costs which are taken against the average wage because they are paid out based on the raw wages of your people. They are calculated as a percentage of the average wage because this is how you will end up paying for them. The more labor on a job, the more of these costs you will incur.

The illustration in the back of this section (page 460) is the form you will use to calculate your labor burden. You will notice there are two columns. One is for field and the other for office. This is because these percentages will be very different. The reason is Workman's Compensation insurance is much lower for office personnel than it is for field personnel. Also, there will not be any figures for vacation, holiday pay, or health insurance: Those costs are calculated in overhead.

The first item is for the company share of FICA. As of this writing, that percentage is 0.077. Notice where the decimal is placed, because that is very important. We are wanting to arrive at a percentage of 100 percent so the percent should not be stated as 7.7 percent.

The next item is your Workman's Compensation insurance. This percentage is based on the stated percentage for the categories of work your people perform. If they perform different kinds of work with different percentages, the percentage should be averaged based on how many of your people do each kind of work. This percentage should either be lowered or raised depending on your modifier. If your percentage was 0.12 percent, but you have not had many claims and they have given you an 80-percent modifier, then 80 percent of 0.12 percent would bring it down to 0.096 percent. However, if you had several claims, your modifier might be 125 percent. You would then take 125 percent of the 0.12 percent to have a new percentage of 0.15 percent. Office personnel would be around 0.005 percent.

The next two items are your federal and state unemployment rates. These are stated as a percentage of your payroll. Whoever does your payroll should know these percentages. They are usually applied only to the first $7,500 in payroll. However, I fix them at their rates because you fire most people before they get to that limit.

Vacation must be calculated using the form provided. It only applies to field personnel. The total vacation weeks are the total vacation weeks of everyone in the field who gets a vacation. For example, if I have 20 people working for me, but only four get vacations, two people get two weeks and two people get one week. My total vacation weeks are six. I divide that by the total weeks worked by everyone. If my 20 people work 45 weeks, then my total workforce weeks are 900. The six vacation weeks divided by 900 workforce weeks would give me a percentage of 0.007.

The next item is health insurance. This item only applies to field personnel. If you don't pay it for field personnel, this becomes a zero. But, if 15 of my 20 field employees get half of their policy paid—some married and some are single—my total bill is $1500 for those 15 employees for a month. That is what I put on the form. I divide that by my total field payroll for the month in which we are working. If my average payroll during a season is $28,000, then I divide the $1500 by $28,000 to come up with 0.054 percent.

The last item shown is holiday pay. Again, this only applies to field personnel. If I pay four paid holidays to 15 people whom I lay off at the end of the season, and six paid holidays to five people I keep all year, then my total paid vacation days for the year are 90. If my total work-force weeks are 900 and my employees work five days a week, then my total work-force days are 4,500. I divide the 90 paid holidays by the 4,500 work-force days to come up with 0.02 percent.

Other items that be included, if you pay them, are sick days, union dues, and any other benefits that apply to labor. I am often asked about bonuses. I do not consider them a benefit which should be guaranteed and paid. Bonuses are paid out of profit, and are only paid if the company is making money.

I have seen labor burden between 0.26 and 0.50 percent, with the average being around 0.30 percent. Office labor burden has been between 0.15 and 0.20 percent, with the average at 0.17 percent.

LABOR BURDEN

	Field	**Office**
Company Share - F.I.C.A.	_____	_____
Workmens Compensation	_____	_____
Federal Unemployment	_____	_____
State Unemployment	_____	_____
Liability Insurance	_____	_____
Vacation (1)	_____	_____
Health Insurance (2)	_____	_____
Holiday Pay (3)	_____	_____
Breaks	_____	_____
_____	_____	_____
_____	_____	_____
TOTAL	_____	_____

1. **Vacation Calculation**
   Total vacation weeks _____ divided by total work force weeks
   in a year _____ = _____.

2. **Health Insurance Calculation**
   Total health benefits payment for a month $_____ divided by
   the total payroll for a month $_____ = $_____.

3. **Holiday Pay Calculation**
   Total paid holiday days _____ divided by total work force days
   in a year _____ = _____.

Labor Burden Form

Overhead Recovery

I equate the running of a contracting business to the flying of an F-14 jet fighter through the Grand Canyon. You drop into the canyon at Point A, which is January second of any business year. You fly 300 feet above the canyon floor with the walls of the canyon 1,000 feet off each wing. You have to fly in a perfectly straight line to Point B, which is December 31st of any business year. You want to arrive with a profit—preferably a hefty one.

On the floorboard of your plane are two pedals. You must keep the pedals even with each other. If one pedal gets out of line, the plane will begin to "yaw" and turn into one of the canyon walls.

"So, what are these two pedals in the jet fighter called my business?" you ask.

One is the amount of overhead you must recover in a business year. The other pedal is the amount of business you must do in a business year to have enough money to recover the overhead, and to make a profit besides. Let's say your overhead is $100,000. You have determined that you must do $400,000 in sales to recover the overhead and make a profit. You drop into the canyon on January second, but by May 15 you see you are behind your sales goal and you are not going to do $400,000 in sales. You will be lucky to do $350,000. Your pedals are now out of line. What is your plane doing? It is yawing. And, if it continues to yaw, you will crash your business into one of the walls of the Grand Canyon.

Now, if you were really in the Grand Canyon and your F-14 plane began to yaw, what would you do in a hurry? You would either push the left pedal forward or let off the right pedal until

Overhead Recovery   461

the pedals lined up perfectly. Do the same with your business. Push the sales pedal forward, or let up on the overhead pedal.

Overhead is a fixed cost. It does not rise and fall based on what you charge for a certain job. You do not need to find out what others charge for overhead and charge the same. Overhead is not charged, it is recovered. And, you do not make any profit until you recover your overhead. To do this you must budget overhead, spend on overhead within this budget, allocate overhead properly, and recover it. In this section—and the following two sections—we'll talk about controlling the overhead pedal.

I know a company which has an accountant who is very nice, except on January 2. She comes in half an hour before everyone else and, with a wicked grin on her face, she turns on her computer and puts an invoice into the printer. She prints out a bill to the management of the construction company for the amount of overhead they must recover in that business year, and leaves a copy on each person's desk.

When the staff comes in reeling from New Year's festivities and football, they are faced with this invoice. They must realize quickly that their plane has taken off and is descending into the Grand Canyon. They know they'd better keep their pedals even so they can arrive safely at the other end of the canyon on December 31st of the new year—with a profit. The staff knows where they stand, and they had better go out the comming year and get enough work with enough overhead allocated on it to recover this overhead, and make a profit.

I recommend you keep track of the number of overhead dollars you must recover in a year in a computer spreadsheet program or on a piece of paper. Every time you sign a contract for a new job, subtract the amount of overhead you have built into this bid from the year's total. Make sure within a business year you do enough work to bring the overhead dollars figure down to zero.

The staff of the construction company with the good accountant does not even talk about profit from January through mid-November because they consider there is none. Any money collected above what it costs to do a job goes to pay overhead. The accountant keeps track and when the overhead is recovered, she puts a copy of the original invoice on everyone's desk stamped "PAID IN FULL." At this point, they have a party. A Break-Even Party. Right after the party, they put the pedal to the metal, because everything else they collect in excess of costs for the rest of the year will be the company's profit.

Here is a point you must understand: You can make money on every job you perform but still lose money as a contractor. Some contractors do not understand what job costing really is. Job costing takes the original estimate of costs to do a job and compares it with the actual costs to do the job. This is done at the level of costs: This is why it's called job costing. Some contractors take their job costing to another level. They add overhead to costs and subtract costs with overhead from the contract amount to see if they have made a profit. That is no longer job costing, it is generating a financial statement or a profit-and-loss statement on each job.

Now, I don't mind if a contractor does that, but he must understand the basic principle of job costing. You see, each of those financial statements on individual jobs are just snap shots. The company and its profitability are a mural made up of those pictures. Individual pictures may look pretty, but the company mural can look real ugly. Why would it look ugly? Because there are not enough pictures in the mural with enough overhead allocated in them to cover all the company's overhead. In order to fill the holes of unrecovered overhead, we must take profit from jobs you did to pay for the unrecovered overhead. This is why you can make money on every job you perform, but still lose money as a contractor.

Why not just add more overhead to the work the company's getting? Because there is a fine line of the right amount of overhead for the kind of company doing certain kinds of work in certain areas. The goal is to find the fine line which both gets you work and recovers your overhead. One of the things our consulting company has done for thousands of companies is to help them find this fine line. In the next section we'll explore how to do it.

For now, think of it this way: contractors make all their profit in four to six weeks. From January through mid-November they are just recovering overhead. Sometime in mid-November they start making profit—everything they do in the last four to six weeks of the year is their profit for the year.

I had a contractor call me just before Christmas one year. I think he waited until then on purpose. His company had broken even the first of October, which left them eight weeks in their business year. He said they billed more work during those eight weeks than any other eight-week period of the whole year and drove their profit into the high six figures.

I always get a sick feeling in autumn and around the holidays—and it's not because of the holiday goodies. First, there is hunting season. Then they are playing Christmas music down at the mall. It's the holiday season and contractors put their planes on autopilot. At the very time they are making a profit, they slow their efforts. I get sick when I think of all the tens of thousands of contractors who have worked very hard to recover their overhead and are finally making a profit, but who coast out the business year just when they could be getting ahead. Remember this next year when you're flying through October toward December 31st!

Budgeting Overhead

When I talk about overhead recovery with regard to estimating I am not talking about the overhead your accountant shows on your financial statement. Accountants are really historians: They are keepers of the historical financial records of your company. They talk and work in the past tense.

As a contractor, you are a futurist. You must put a price on a future job with your future overhead built into the price so, when you are in your future doing the future job, you will recover your future overhead. Did I get the word future into this concept enough? You are a futurist! Consequently, you cannot use the overhead your accountant gives you now or you will be going out into your future and recovering the overhead of your past, a practice which will definitely put a damper on your future success.

You need to use your accountant's financial records as a reference to do your budgeting. You will look at what you spent last year for an area and ask yourself if the amount of increased or decreased business you intend to do this year will lower or raise the costs from the previous year. Sometimes you can increase your business without increasing overhead in certain areas. Sometimes a decrease in business will not decrease all overhead items because of the decrease in sales. Each item must be considered in this light. Also, certain items may be increasing because of inflation. When you fill in next year's budget for each item you will need to be a futurist and project what next year's cost will be based on these considerations.

The other reason this overhead budget is not based on your accountant's overhead is your accountant has included things in overhead which we have put in equipment costs or labor burden. Some examples are equipment depreciation, interest on equipment loans, maintenance of equipment, licenses, fuel, insurance on equipment, payroll taxes, and payroll insurance. Many times these items are in overhead as your accountant has defined overhead. However, we put those costs either in the daily or hourly cost of your equipment or into your labor burden percentage.

There are equipment costs in overhead, the cost of the equipment used by overhead people. And, there are labor burden costs on overhead salaries, which you calculated as an office labor burden percentage. These items will be added into the overhead budget based on your calculations rather than on the accountant's figures. The important thing to focus on when you are going through your financial statement and using it to make a budget is these things will be taken out of your accountant's overhead and your overhead. The section on page 277 of this book is dedicated to financial formats we would recommend your accountant using to make budgeting easier, and to give you the ability to compare your budget to your actual finances.

In the back of this section (page 473) is a form we want you to use to budget your overhead. We suggest you prepare a form like this to work on in pencil, because you will be making more than one change to your figures as you work through them. Before you start, we must talk about one more consideration in the budgeting process. If you have more than one division (what some people call a profit center), you will need to come up with an overhead budget for each division.

Let's first define what a true division is and is not. I recommend contractors set up a division only when the divisions are doing things which are dissimilar. For example, a Landscape Contractor would not set up a landscape division and an irrigation division. Landscaping and irrigating are similar. They are both contracting, and they both require you to send out material, labor, and equipment to do a job.

There are four true divisions in this kind of business which are dissimilar. One division is contracting of all kinds, another is maintenance, which is dissimilar from contracting because it is more service-oriented. It is repetitive work with very little material but high labor and equipment. Another division is a nursery operation. This is a type of farming that requires a period of years in order to harvest a crop. It is dissimilar from both contracting and maintenance. The other true division is a garden center. This is a retail operation which requires sales clerks, advertising, and store layout. It is dissimilar from contracting, maintenance, and nursery operation.

For each division you have you must create a separate budget. In overhead, it will be very obvious where some items go and how much of various costs should go into each division. But some items, such as telephone costs or office supplies, are more difficult. If you can separate the costs, do so. If it is too difficult, we recommend you use what we call the "ABC Method." A stands for sales, B stands for overhead people, and C stands for field people.

Go through each item in the overhead budget and ask this question: Is this item affected by sales, overhead people, or field people. Then put an A,B, or C by that item.

You will now need to come up with a percentage split for each of the A, B, C categories. For example, the category A item of sales. Let's say your sales for a year for the entire company is $1,000,000. Of this total, $500,000 is in contracting, $300,000 is maintenance, and $200,000 is in some other operation. This would mean you divide all A items at the following percentages; 50 percent to the contracting division, 30 percent to the maintenance division and 20 percent to the other. You arrive at the same kinds of percentage breakdown for the B category of overhead people by discovering what percentage of time each overhead person spends working on the different divisions' concerns.

You then take your total field people and arrive at a percentage of how many work in each division in relationship to the total people in the field. Finally, you arrive at your percentage for the C category of field people. You apply these percentages to the total amount of money spent in an overhead category to arrive at how much of that total would be budgeted to each division. We will now go through each overhead category and explain what is and is not included in each one.

◆ ADVERTISING:

    a ···Yellow page ads
    b ···Newspaper ads/ help wanted ads
    c ···Magazine ads
    d ···Door hangers
    e ···Bill boards
    f ····Bulk mail
    g ···Brochures (prorated over the number of years you will use these)
    h ···Garden show costs

Does not include business cards.

◆ DEPRECIATION (Office equipment and furniture):

(Take the total cost of each item and divide it by the years you can expect to use it).

    a ···Desks
    b ···File cabinets
    c ···Copy machine
    d ···Blue print machine
    e ···Calculators
    f ····Telephone system
    g ···Plan rack
    h ···Drafting table
    i ····Typewriter
    j ····Fax machine
    k ···Plants, pictures, and miscellaneous

Does not include field equipment, computers or radios,
    or cellular phones.

◆ DONATIONS:

    a ···Any cash or time donation that will put you in a good light in the public eye and get you business. Does not include donations from profit to reduce your tax burden.

◆ DUES AND SUBSCRIPTIONS:

    a ···National organization dues
    b ···Local organization dues
    c ···Business magazines
    d ···Bid reports
    e ···Plan room dues
    f····Warehouse club memberships

◆ INSURANCE:

    a ···Contents policy on office equipment
    b ···Health insurance for overhead people
    c ···Life/Key-man insurance, as long as company is the beneficiary

Does not include insurance on equipment, payroll or Workman's Comp.

◆ INTEREST AND BANK CHARGES:

    a ···Interest on line of credit (If you don't borrow from bank, amount you are loaning the company with retained earnings.)
    b ···Interest on charge cards
    c ···Supplier interest
    e ···Bank service charge fees
    f····Bounced check fees
    g ···MasterCard and Visa fees

◆ DOWNTIME:

This is for nonbillable time of field people as follows:
    a ···Equipment or truck breakdown
    b ···Shop work excluding equipment repairs time
    c ···Work on office or property

This should be calculated in hours multiplied by the average wage.

◆ LABOR BURDEN (DOWNTIME):

Multiply your downtime figure by your field labor burden percentage.

◆ OFFICE SUPPLIES:

a ···Pens, pencils, paper
b ···Stationery, invoices, business cards
c ···Postage
d ···Small pieces of office equipment

◆ PROFESSIONAL FEES:

a ···Outside CPA
b ···Outside bookkeeper
c ···Legal fees
d ···Consultants
e ···Payroll service
f····Cleaning service

◆ RENT:

a ···Your rent payment
b ···If you own building, fair market rent value for like properties
c ···Storage facilities or satellite yard

◆ SALARIES–OFFICE:

a ···Secretary
b ···Accountant
c ···Designer/estimator
d ···Field supervisor (prorate if supervisor works on a job part time)

◆ SALARIES–OWNER:

a ···Owner(s) (prorate if owner spends time working in the field)
___

The total of these two should not exceed 45 percent of the total overhead.

◆ LABOR BURDEN (OFFICE)

Multiply your total office and officer salaries by your office labor burden percentage.

◆ SMALL TOOLS AND SUPPLIES:

a ···Wheelbarrows

b ···Shovels, rakes, maddox, picks, etc.

c ···Hand tools

d ···Repair parts for small tools

e ···Nuts, bolts, nails

◆ TAXES–BUSINESS

a ···Business privilege taxes

b ···Yearly taxes on office equipment

———

Does not include sales tax on materials

◆ TELEPHONE:

a ···Line charges

b ···Long distance charges

c ···Rental of system

d ···Answering service

e ···Repairs

◆ TRAVEL AND ENTERTAINMENT:

a ···Travel costs to seminars, conventions, tree tagging

b ···Meals for clients and architects

c ···Christmas or Thanksgiving party

d ···Summer picnic

e ···Hotel bills

◆ UTILITIES:

a ···Sewer charge

b ···Water charge

c ···Electric charge

d ···Gas and oil

e ···Trash service/non-job-related dump cost/dumpster fees

f ····Office cleaning service

◆ YARD EXPENSE:

Cost associated with maintaining a storage yard. And leasehold improvements to your office depreciated over the life of the improvements.

◆ OVERHEAD VEHICLES:

a ···Vehicles used by overhead personnel for overhead purposes. (Trucks that don't go out on the job).

b ··· Mileage reimbursement for sales peoples' vehicles.

◆ RADIO SYSTEMS/CELLULAR PHONE

   a ···Two-way radio system (Depreciate over 5-10 years/useful lifetime.)

   b ···Beepers

   c ···Line charge

   d ···Repeater charge

   e ···Monthly maintenance

   g ···Service

   h ···Car phones (Depreciate over 2-3 years.)

◆ MISCELLANEOUS: Limited. Keep under $500.

◆ LICENSE AND BONDS:

   a ···Paper license for the state you operate in

   b ···Bond-license bond only

   c ···City licenses if applicable

◆ EDUCATION:

   a ···Workshops

   b ···Seminars

   c ···Community College

   d ···Conventions

   e ···Registration for conventions and trade shows

◆ UNIFORMS:

   a ···Shirts

   b ···Cleaning service

   c ···Safety equipment; i.e., goggles, gloves, ear protectors, hard hats

◆ COMPUTER AND HIGHLY SOPHISTICATED SYSTEMS:

   a ···Computer, software, and digitizers

   b ···Depreciate over a three-year period

   c ···Include costs to have personal computer consultants train your staff

◆ BAD DEBTS:

   a ···Usually 1/2 of 1 percent of gross sales

   b ···Percent of accounts receivable that are written off

LIST OF OVERHEAD ITEMS
FOR BUDGET PURPOSES

Advertising	$ _____
Depreciation (office equipment & furniture)	$ _____
Donations	$ _____
Dues and subscriptions	$ _____
Insurance (office items and health/life)	$ _____
Interest and bank charges	$ _____
Downtime	$ _____
Labor burden (downtime)	$ _____
Office supplies	$ _____
Professional fees	$ _____
Rent	$ _____
Salaries-office	$ _____
Salaries-officer	$ _____
Labor burden (office)	$ _____
Small tools and supplies	$ _____
Taxes-Business	$ _____
Telephone	$ _____
Travel and entertainment`	$ _____
Utilities	$ _____
Yard expense	$ _____
Overhead vehicles	$ _____
Radio systems	$ _____
Miscellaneous	$ _____
Licenses, bonds	$ _____
Education	$ _____
Uniforms and hard hats	$ _____
Computer	$ _____
Bad debts	$ _____
Total Overhead	$ _____

Different Methods To Use To Allocate Overhead

After you have budgeted your overhead for the next year, you must determine how you are going to allocate it to your different jobs. If you have an overhead of $100,000 to put on $400,000 worth of work, how are you going to do it? There are three different ways: "Close But No Cigar"; "Deadly"; and, "As Close As She Gets."

Close But No Cigar. This concept puts all your overhead on labor. If you had 10,000 hours worked by your field people you would divide that into the total $100,000 overhead to come up with $10 per hour. You would then add that figure to every hour on your bid. This concept works well for maintenance contractors, but for various building contractors, it is "close but no cigar."

This method says that labor is responsible for all of your overhead. It says that material does not burden or tax your overhead. This is not true. Material burdens overhead because someone needs to take it off of the drawings, price it, buy it, get it delivered, check all the invoices, and get the suppliers paid. Equipment burdens overhead when someone needs to buy it, get it serviced, move it from job to job, and make the payments on it. Subcontractors burden overhead when someone needs to call them, meet them to go over their work, write a contract, coordinate them, call their phone machine again and again and again, and pay them.

As Close As She Gets. Each item of a job burdens or taxes your overhead and must pay its portion of their costs. That's why we believe all types of contractors must use a multiple-overhead-recovery percentage system, which allocates some of the overhead on each of these items. This system is "as close as she gets." There is no perfect way to allocate overhead, and if anyone tells you there is tell them they're lying. I believe in a multiple-overhead-recovery percentage system based on the fact that labor is the greatest cost of overhead. Let me illustrate.

I could go into business tomorrow selling contractors material for their jobs. I could get an 800 telephone number in my office at home and buy a nice oak desk and a leather swivel chair. Contractors from all over the country would call me for prices on my material which they have specified on jobs. I would quote them prices FOB job site. If they got the job they would call me and order the material. I would call the factory, arrange for trucking, bill the contractor, and hopefully, get paid. If I sold $300,000 worth of material a month and charged 5 percent overhead, I would get $15,000 a month for overhead. I could pay my telephone bill, my house payment and utilities, buy a new desk twice a year, and still make a good monthly salary.

However, what happens to me if I start installing that material? I will need a secretary to answer all the telephone calls from people complaining about me not being there on time or the work being done wrong. I will need a big office to house my expensive estimator, designers, accountant, and field superintendent. I will need an accountant to do all the payroll and job costing and to keep track of all the other payables and receivables. I will need an estimator or designer to go get more work and a field superintendent to see to it that all the work gets done. In other words, I will have overhead in a big way. Why? Not because of the material but because I was installing that material. Labor is not only the greatest risk in the contracting business, it is the greatest cost of overhead.

Deadly. In a multiple-overhead-recovery system you will have four different percentages to add to your material, labor, equipment, and subcontractors. Of the four numbers, labor is the highest, equipment the next highest, then material, with the subcontractor number being the lowest. Let me compare this system with what we call the "deadly" system." The deadly system is one percentage applied to all costs equally. If you had an overhead of $100,000 to be added to $275,000 in costs, you would add a figure of 36 percent overhead.

Let me illustrate why that is a "deadly" system. Let's say we are both bidding two different jobs. The first one has the following total costs:

Material	Labor W/Burden	Equipment
$100,000	$10,000	$10,000

The second one is a labor-intensive job and has only a little bit of material. It is a remodel job or the owner is buying the material and it has the following total costs:

Material	Labor W/Burden	Equipment
$10,000	$100,000	$10,000

Both have a total cost of $120,000 to which you add 36 percent for overhead or $43,200. However, I add 10 percent overhead to my material, 60 percent to my labor and 25 percent to my equipment, because I believe that because labor is the greatest cost of my overhead it should recover the greatest cost. I would then add $18,500 overhead to the first job in comparison to your $43,200. On the second job I would add $63,500 in overhead compared to your $36,000. Guess which job I get? Guess which job you get? Guess which job I'm glad you have because you won't be around for very long. Guess which job you will wonder how I can do so cheaply.

But, look, if my payroll is $10,000 a month, I will be in and out of that first job in one month while you are on the second one for ten months. Overhead for one month is a lot less than overhead for ten months on the second job.

"Well," you say, "That's a far-out example!"

Yes, it is, but I have seen them like that. There is a ratio which it is very important for you to understand. It is called the material, labor, equipment, subcontractor ratio. And on every job it is different. Sometimes the material is very high, the labor is low, and so is the equipment. Sometimes the material and labor are equal and the equipment is high. Sometimes the material is really low, the labor really high, and the equipment is in the middle. It changes with every job you will ever bid. Consequently, your overhead recovery numbers will change on every job.

Now, let's attempt to arrive at those four percentages. This system I have found is the most understandable and reasonable system in its application I have ever used. Finding your overhead recovery numbers is not a slam-bam event; rather, it is a process. You come up with the numbers and then sample them in the marketplace. You fine tune them and fine tune them until you have percentages which are comfortable.

You take the total overhead which is to be recovered for your next year from the previous section. You then build a year into one job. This can be done by using a form similar to the illustration in the back of this section (page 482). If you have a previous financial statement which is done in such a way that you can pull off the material, labor and subcontractors from the previous year, this is your best approach. Then decide how much you will increase or decrease your labor force in the field based on your overhead, key people, and the ability to get work. Labor is the big risk, and the largest overhead requirement, so I feel labor is the cornerstone of our strategy. If you think you will add 10 more people to your current 50, you will have an increase of 20 percent, or vice versa. Take last year's labor and material and add or subtract 20 percent. Insert those figures on the form under the proper heading. If the material did not include tax, add tax to it before you bring it down to the total line. Then add your field labor burden to the labor and bring down the total labor with labor burden.

Then you must decide about subcontractors. If you make very little use of subcontractors, and there are no changes in your approach to your jobs, increase last year's figures by 20 percent. If there are changes, try to anticipate those changes and the final number of subcontracting dollars you will spend.

Next is the equipment figure. If you have kept track of equipment from time cards, arrive at the equipment cost based on hours worked times the figures you use in the bid process, then you will only need to add 20 percent to this figure and use it. This process is further explained in the section on equipment (page 443). If you did not keep your books in this way, you can take your equipment, estimate its hours of work last year, multiply those hours times each piece's hourly price, and arrive at a total figure to which you add 20 percent. You should now have a reasonable material, labor, equipment, and subcontractor cost for the next year, anticipating growth or regression.

If you do not have a good set of records to use, you must arrive at your figures using a less-accurate method. Again, labor is the cornerstone. It is arrived at by taking your anticipated work force times the hours you feel they will work. In doing this you need to take into consideration bad weather days, slow seasons, and busy seasons with 50- or 60-hour weeks. Once you have arrived at a total number of hours, multiply them by your average wage to come up with your labor figure.

To arrive at your material figure, check the material (with tax) to labor (without labor burden) ratio of some recent bids. Let's say in totaling up the material and labor on five or six bids you discover material is $225,000 and the labor is $100,000. The material to labor ratio is 2.25 to 1, labor always being one. Now take this ratio to your overhead recovery calculation sheet. Multiply your labor calculation by the material ratio to come up with your anticipated material purchases for a year. The equipment and subcontractors would be arrived at by the same method.

Now you should have two important sets of figures. The anticipated costs for next year, and the overhead you need to recover from those costs. From here I use an approach which is quite simple. I use three fixed percentages and find the fourth, it being whatever the three do not recover. From our study of this industry, we have established three of those figures which are the normal or average burden which is put on overhead by these items. Here are the three fixed percentages we suggest:

Material .... 10 %
Equipment .... 25 %
Subcontractors .... 5 %

You apply those three percentages to your anticipated costs and find out how much of your overhead will be recovered. Whatever isn't recovered is the amount which *must* be recovered off of the labor. You divide the remaining amount of overhead which was not recovered through the application of those three percentages by the amount of your anticipated labor with labor burden to arrive at the percentage to be used against your labor on bids. I have seen this figure as low as 30 percent and as high as 95 percent, depending on the contractor or seasonal conditions.

Now you can apply those numbers to your anticipated costs, add profit, and you have a goal of how much work you must accomplish in order to recover your overhead. If you miss the goal, cut back overhead. If you want to add overhead you will quickly be able to carry it through the system and see how much more work you must do or how much it will raise your overhead recovery markup percentages.

One thing you must take into consideration in watching all of this is the amount of work you subcontract. If it varies from your original estimate, your work-accomplished goal needs to vary accordingly. If you sub more, increase your goals, and vice versa.

The work we have accomplished is for a contractor who does one kind of work, with all their jobs being about the same size. However, a great many of you don't fit into this category. There is one more variable which must be taken into consideration; the size of the job. The smaller the job, the more overhead it requires and vice versa.

For example, if you are a contractor who does large commercial or public works projects and residential or smaller projects, plus you do service or maintenance-type work, then the smaller jobs require, per dollar spent, more of your overhead.

This is my suggested approach. First, break down your company into divisions. For example, a large project division or approach and a residential division or approach. Then appropriate your overhead, item by item, to each division. You will then come up with the amount of overhead each division must recover. Then prepare a different overhead recovery sheet for each and arrive at your overhead percentages as previously directed.

Another situation which will require you to set up a different set of overhead recovery numbers is when you are doing both low-wage, and union or federal wage jobs. Again, you will need to set up the previously mentioned numbers based on a year of doing all your work with each of the two kinds of pay scale. Because the federal and union jobs will have a higher labor cost, your material-to-labor ratio will need to be refigured. You will also find your overhead recovery number for labor on those jobs will be less than your nonunion or low-wage jobs. This is due to the fact you are recovering the same overhead off of a higher labor dollar, but your risk has not increased.

I have seen a lot of overhead recovery systems, and every one of them requires you to do your homework in determining how well they work in the marketplace. If a system is pricing you out of projects, you need to reduce your overhead, and plan on doing more work.

OVERHEAD RECOVERY MARK-UP CALCULATIONS

If your costs are. . . . .$_____

And your overhead is . . . .$_____

With a profit of . . . .$_____

Your sales will be . . . .

IF

Material with tax is	*Labor is	**Equipment Is	Subs Are
_____	_____	_____	_____

Add Labor Burden
@_____%   _____

_____

$_____ $_____ $_____ $_____

Then you overhead recovery will be. . .

Material :_____

Labor   :_____

Equipment:_____

Subs   :_____

TOTAL   :_____

* _____people @_____ hours = _____hours x_____average
   wage =$_____

** Equipment calculated on estimated hours of your equipment
   times its cost per hour

Profit

There are two concepts about profit which contractors must understand. First, profit is a payment for risk. The profit should go to those who take the risk of losing their investment and even their personal wealth. It is a reward to those who are willing to take those risks.

Second, profit is a return on investment. If you put your money in a bank or invest in mutual funds or stocks, you can expect interest. (You expect your stocks and mutual funds to increase in value.) Profit in a company is the return on the retained earnings and the initial capital which was put into the company.

The combination of these two concepts in contracting comes together as follows:

A balanced portfolio of money market accounts, certificates of deposit, some mutual funds and stock investments is returning an average of 9 to 11 percent on investment. However, such investment are fairly low-risk. Because construction is a high-risk business, the payment for that risk should be two to three times as much as the lower risk investments. This means that construction companies, as a return on investment and a payment for risk should have a 20 to 30 percent return on retained earnings and original capital.

With these thoughts in mind, I would encourage you to look at the last couple of years' profit as it relates to your investment.

Before you can measure your profit, however, you must know what true profit is or is not. Let me give you some examples of how contractors deceive themselves about the nature of true profit. Contractors can show an excessively large profit for any or all of the following reasons:

- The owner does not take a salary, only draws on profit which are shown only on the balance sheet, or a conservative below-market-value salary for themselves.

- Family members (usually the wife) work for the company without receiving a salary.

- The company offices are in the home, and because of IRS rulings, nothing is paid by the company as rent.

- Old equipment is being used by the company, which has a zero net worth because it has been depreciated. The company is not setting aside expense money for future replacement equipment.

In order to determine a "true" profit, these shortfalls in expenses must be deducted from the existing profit. Contractors can show an excessively low profit for any or all of the following reasons:

- The owner(s) is taking a larger-than-market value salary for themselves. They are taking profit out of the company as salary.

- They are building a house or remodeling a house and writing off some of the expenses through the company. Or, they are maintaining their, or a relative's, house by doing the same thing.

- They have started another company or profit center in their existing company and they are getting the startup costs out of profit.

- They have made a large contribution or taken an expensive trip (business, of course) and the company has expensed it out.

- They made some large purchase of computers or other equipment that has been expensed.

The total of these items, if they exceed a reasonable cost in a given year, must be added back into profit in order to determine a "true profit."

I suggest that, in order to establish your profit markup on a bid, you begin by having a profit range. Establish the lowest profit you would accept. For example, let's say the low range is 10 to 15 percent, and the high range is 25 to 30 percent. Then you must consider four things about the job and apply a profit figure to each of the four considerations between the low and high as if each item is the only thing you are considering. After you have done this, then add up the four profit percentages, divide this number by four, and you will have an average profit percentage for the job, based on your considerations. I like this systematic method better than adding 12 percent profit to a job and, saying when asked why, "On 'Good Morning America' this morning said there was a 12 percent chance of rain, and it sounded like a good profit figure to me." Let me tell you what those four considerations are:

Need. That is, how badly do you need the job? The greater the need, the lower the profit percentage should be. But, if you have a lot of work, the profit percentage should be toward the higher end of the range.

Risk. Risk comes in several different ways. Labor is the biggest item of risk in construction. Therefore, the more labor there is in a job in relationship to material costs, the greater the risk. If you do not feel comfortable with the number of hours you have estimated to do the project, then your estimate becomes a risk. If there are unusual or potentially troublesome conditions on the site, e.g., rock, water, poor accessibility, etc., then the job is a greater risk. The more risky the job, the greater the profit you should seek. If it is a "piece-of-cake" job anybody can do, then anybody and everybody will want to bid it and, hence, the profit will have to be low. But, if it is very difficult, very few will be dumb enough to bid it—and those who do had better go high on the profit because of the risk involved.

Size. That is, size in relationship to what you are accustomed to doing and the size of the jobs upon which you have based your strategy. Let's say that your best size project is $30,000 and along comes a job for $50,000. Go lower on the profit. If the job is only $8,000, go higher on the profit.

Marketplace. This is the one we all like. The question here is, how many people are bidding the job and who are they? Is it "The Good Times," with lots of work for everyone, or "The Lean Times," when contractors are cutting each other's throats? The more bidders there are or the tougher the times, the lower the profit you add.

The Ding-Dong Factor

I remember bidding a site-development job once. It bid at 2 P.M., and because there was no electrical work on the job, I was finished with it by 10 A.M. It was all filled out and sitting in an envelope on the edge of my desk. My bid was $838,000.

At around 11 A.M., I got a call from one of my good friends, my bonding agent.

"Charlie," he said, "I see you've got a bond for such and such project. Have you ever worked for this architect, Mr. Ding-Dong, before?"

"No, I haven't," was my response.

"Well, let me tell you a few things." And, he went on to say, there is only one other contractor who has even considered pulling a bond on this job. Furthermore, this architect has gotten his reputation by telling clients that he will cost the contractor money. He will get them to bid a job thinking they know what's included, but he will get them to do more work than they planned on. Furthermore, he has a broken accent, which he uses to throw off any discussion. His favorite statement is, "No shange order my jobs! no shange order my jobs." It doesn't matter if you can't find a reference to what he wants you to do, he'll say, "No shange orders my jobs!" He'll tell you it was on another set of plans for another job he did and you should have known this when we bid the job.

I should have been tipped off, because this $838,000 job had only three sheets of drawings. My six-year-old daughter printed better than someone printed on those plans. It looked like some of the words were printed on the original in pencil and then smudged.

I said thank you to my good friend the bonding agent, and hung up the phone. I took out my bid proposal and changed a 3 to an 8 and bid the job at $888,000. I added $50,000 for the ding-dong factor. The only other bidder came in at $900,000-plus and we were awarded a contract.

When I set the job up for the computer, I actually had a cost code for the "Ding-Dong Factor." I put $30,000 under that category (I wanted to make $20,000 off of the ding-dong) and told my project superintendent that every time Mr. Ding-Dong came out there and asked him to do things that weren't anywhere on the plans that he should let me know. We'd submit a change order, only to hear "No shange orders my job." We'd send letters, argue and fight, but finally we'd charge it off to the Ding-Dong Factor.

When the job was all done, I went out to do the punch list. The architect was there with a big grin on his face and an attitude which said, 'I cost you money.' I took on a down-in-the-mouth defeated look during the whole meeting. But, as I was driving back to my office, I was grinning like a bear. Before I left my office, I pulled up a report which told me only a little over $25,000 had been charged to the Ding-Dong Factor.

Now, I know ding-dongs in Denver, Dallas, Los Angeles, and San Diego. I know ding-dongs in Houston, Kansas City, and Atlanta. Sometimes they are architects, sometimes they are owners, sometimes they are developers, and many times they are general contractors. They are people who take advantage of other people. They squeeze them for all they can get. They pay slow and demand much. They shop your bid and expect deals all the time. They never admit they are wrong, yet know you are always wrong. They are ding-dongs.

We have to police this industry ourselves. The only way I know of doing this is through the Ding-Dong Factor. It doesn't take long for these ding-dongs to get a reputation and for owners to find out to use these characters costs them more money on the bids. General contractors lose bids because their subcontractors are bidding the work higher because they don't want to work for these people.

I remember bidding a job to a general contractor once when I was working for a subcontracting firm. After the bid opening, I received a call from the low general contractor's estimator. He wanted to have a meeting with me the next morning. I asked if I was the low number for my type of work and he assured me that I was, but said they needed to talk with me.

The next morning I went to the general contractor's office. I was ushered into the estimator's office. He was sitting at his desk with his chin in his hand. His superior was standing, leaning against the wall.

I said, "Good morning. We got a job, huh?"

"Charlie," was the reply, "We had a bust in our bid that's going to cost some money. We are calling in all the subcontractors to see if they can shave a couple of percent off their bids. It'll help spread around the problem over several contractors." I was waiting for violin music to come over the Musac system. They were trying to squeeze me.

I stood up and said, "Gentleman, I'm a big boy and you're big boys. I live with my numbers and so should you." And, I left. At every opportunity, I talked with other subcontractors in our trade so we all began to add the Ding-Dong Factor to our bids to this kind of general contractor. You see, what goes up, must come down, and what goes around will eventually come home.

I have a code I live by when it comes to shopping prices. It goes like this. If I get a price someone did not have to spend any time or money getting for me, I feel I can shop this price. If it came out of a catalog or off a price sheet, I feel I have a right to shop this price. However, when someone spends time or money to get me a price, then I had better honor it and give it to the *best* price which was available *before* the bid opening.

There is a contracting company in this country which I respect very much. They have a policy that states that they will fire any estimator who is caught shopping bids. Subcontractors respect them so much they are standing in line to bid to them.

But let me add one more thought to this concept of the Ding-Dong Factor. I do not add a contingency to every bid. Remember, I have a fudge factor (or contingency) added into my average wage. I don't want to be in a position of adding a contingency to every bid. Then I would be relying on my feelings and my attitude towards the job, and I've never found this to be productive. I want to do my best on estimating a job and put a Ding-Dong Factor on the end when I really know it applies.

A Sample Bid

There is one word which describes the point of the system we are about to show you: simplicity. In the bid process you want to take two or three hundred pages of specifications, plus two or twenty sheets of drawings, and simplify them to a piece of paper which you can work on just before you must turn in your bid. Remember, most estimating mistakes happen in the last hour or two before you turn in a bid. This system will help you reduce those errors.

The first thing an estimator should look at is the proposal form. See how you are to submit a bid. If it is a lump sum bid, read the specifications, decide how to break down your bid; that is, decide what to include in the breakdown. If it is unit price bid you'll want to know where you must include items for which there is no unit price item on the proposal form.

Take other things into consideration before you start. You'll want to know if there are alternate adds or alternate deducts. Most contractors don't mind alternate adds and they estimate them just like change orders added onto a project. Alternate adds tell them this owner and architect are starting conservatively and are hoping to add to the project.

Alternate deducts tell them they may have an owner and architect who don't know what they are doing, or they want a Cadillac at a Volkswagen price. Alternate deducts are more frightening and must be approached carefully. You see, when contractors look at a project they see it as a complete unit. They can visualize the completed unit, estimate its costs and then add more items to it. But, it is difficult to visualize a complete unit and estimate how much certain things will save if you take them off. To complicate things even more, they must estimate without knowing which things the owner or architect may finally decide to deduct. To protect themselves in this situation some contractors estimate the job without deducts, then figure the deducts as adds, and add them to the bid to arrive at the lump sum. Then they give the deducts based on what they figured. That way, if all the deducts are taken, they will end up with their original estimate. However they decide to handle them, they must be careful with alternate deducts.

After you have established the type of proposal the owner wants, you begin to read the specifications and make notes of pertinent items on a legal note pad. You should not highlight specifications. You will find when you do that, you still have to carry the big book around to find all your highlights. And, if you lick your finger too much, a couple of pages could stick together and you would miss some important items. If you make notes on another paper, you can check off those items as you pick them up.

You then go to the plans and familiarize yourself with them. Where are certain details and what sheets can you break out and give to the subcontractors? You then go over every note and detail and again make notes. You initial those notes after you have taken them into consideration.

Having completed your reading and familiarization, you are ready to set up your bid sheets. If it's a lump sum bid, you break the job down into work divisions and make recap sheets. Those items you are doing with your people, you break down into work functions; i.e., earthwork, concrete, carpentry or landscape, irrigation, etc. Then you list all your subcontractors near the bottom or on the recap sheet. (See "The Sample Bid.") This is the single sheet you want to boil everything down to—and the only sheet you want to be working on an hour before the bid.

Then you take additional bid sheets and make a sheet for each item of work you are going to perform per your recap sheet breakdown. It is on these sheets you will do your actual calculations of what material, labor, and equipment will be needed per work item.

Notice on the illustration I figure my wholesale cost of materials per quantity takeoff. I then figure the production hours it will take to install the materials. Then I add any specialized equipment needed to do the work, its hours and costs. You will also notice a general condition sheet has been added so general conditions can be figured as a separate item. This must also be added before you start doing those calculations.

There are a lot of different kinds of estimate sheets. Some list all the items which usually appear on the job. The estimator is then required to go through the job and fill in the blanks of the estimator sheet. That kind of form can make someone a lazy brain. It's estimating by filling in the blanks, and if there is no blank, there is no figure.

This estimate sheet is basic and simple, but it doesn't make life easy for the estimator. This is good, because then an estimator must build the job carefully and systematically in his or her mind and on paper just the way it's going to be built in the field.

If you will look at "The Sample Bid," you will see how to do each separate worksheet. You begin at the beginning of a work function and calculate the material or labor equipment which will be needed to do the item. You can then insert a unit cost for the item and multiply it by your quantity to come up with a total material, labor, or equipment cost. You can then bring a total down so on each work sheet you will arrive at a total material, labor, and equipment for the work function.

Now, once each separate work sheet is completed and totaled, you carry those totals for material, labor, and equipment to your recap sheet. This is done an hour or two before bid time and then you are working on only one sheet just before the bid is due. You have taken numerous pages of specifications, plans and worksheets and simplified them down to one sheet.

It is on this sheet you can add your sales tax to all material, and labor burden to labor. To those totals, you add your overhead recovery numbers per the previous section. Then you can total those numbers and add your profit contingencies.

As you work through the separate worksheets, you are able to total the number of man-hours you are estimating to do this function. You can then divide those hours by your anticipated crew size, and then by the number of hours the crew works in a day. This will then give you some guidelines of how you are going to run the job and how long it should take. If you total all those days together from each individual sheet, you will have the number of days the crews will be on the job. Once you get the same kind of information from your subcontractors on the job, you can have a tool for estimating the general conditions.

For example, if the job will take 35 working days, you will need 35 days of some amount of supervision, 35 days of possible mobilization to the job, 35 days of job clean up, 35 days of pickups, trailers and toilets on the project. This also gives you an instant ability to know if this job is costing more than estimated. The minute the job goes past 35 days, you know you are losing money. Also, before your company ever starts the job, you can tell your people when the job must be completed for your company to avoid losing money.

In summary, you will notice these bid systems have a couple of key ingredients. One, you always give the contractors a total of material, labor, equipment and subcontractors for the job so they can consider this important ratio. Two, the bid systems always deal in production hours so the contractors can predetermine how long they will be on the job, plan accordingly and estimate the general condition accordingly. Third, they can take two to 30 sheets of drawings and two to 300 pages of specifications and simplify it all down to one or two sheets. This should improve their accuracy greatly.

The following is a complete set of estimate sheets for a lump sum bid.

PROJECT
LOCATION
CONTACT
DATE   SUMMARY BY

BID ITEM	QUANTITY	UNIT	PRICE	MATERIAL	LABOR	EQUIPMENT
GENERAL CONDITIONS				840—	2,080—	1,120—
CLEAR AND GRUB						
EARTHWORK						
CONCRETE, STRUCTURAL						
CONCRETE, FLATWORK						
PLAYGROUND						
RESTROOM BUILDING						
STORM DRAINAGE						
IRRIGATION				11,406—	5,510—	2,640—
LANDSCAPE, PLANTING				10,581—	3,090—	410—
SEEDING				7,750—	3,820—	1,220—
MISCELLANEOUS						
TOTALS:						
				• OVERHEAD	• Labor Burden) OVERHEAD	
					• OVERHEAD	
				TOTAL		
				MATERIAL:		
				EQUIPMENT:		
				SUBCONT.:		
					• PROFIT	
					• CONTINGENCY	
					TOTAL BID	
SUBCONTRACTORS:						
ELECTRICAL						
PLUMBING						
ASPHALT						
TOTALS:						
OVERHEAD:						
TOTAL:						

Estimate Recap Sheet

PROJECT _____

LOCATION _____

DATE _____

BID ITEM	QUANTITY	UNIT	PRICE	MATERIAL	LABOR	EQUIPMENT
GENERAL CONDITIONS .						
SUPERVISION = 28 DAYS x 3 Hr	84	Hrs	10 —		840 —	
MOBILIZE = 28 DAYS @ 1 Hr.	28	Hrs.	10 —		280 —	
CLEAN-UP = 28 DAYS @ 2 Hr	56	Hr	10 —		560 —	
HAUL MATERIALS	8	Hrs	10 —		80 —	
PICK-UP TRUCK 28 DAYS @ 8 Hr	224	Hrs	5 —			1120
TOILETS	1½	Mos	60 —	90 —		
TRAILER	1½	Mos	100 —	150 —		
SIGN	1	EA.	250 —	250 —		
LABOR TO SET UP YARD	32	Hrs	10 —		320 —	
TESTS	5	EA.	60 —	300 —		
TEMPORARY WATER	—	LS	—	50 —		
TOTALS				840 —	2080 —	1120

THESE HOURS ARE FICTITIOUS

General Conditions Estimating Sheet

PROJECT _____

LOCATION _____

DATE _____

BID ITEM	QUANTITY	UNIT	PRICE	MATERIAL	LABOR	EQUIPMEN
CONCRETE FLATWORK						
GRADING — LABOR	8	HRS.	$ 10 —		80 —	
GRADING — TRACTOR	4	HRS.	$ 15 —			60 —
FORMS	500	FT	$ .50	250 —		
MESH	3,000	SQ. FT.	$ .10	300 —		
LABOR — FORM — 2 MEN/2 DAYS	32	HRS.	$ 10 —		320 —	
LABOR — INSTALL MESH	8	HRS.	$ 10 —		80 —	
CONCRETE	40	YDS.	$ 60 —	2,400 —		
EXP. JOINT	340	FT.	$ .20	68 —		
MISCELLANEOUS	3,000	SQ. FT.	$ .05	150 —		
LABOR TO POUR = 4 MEN/1 DAY	32	HRS.	$ 10 —		320 —	
LABOR TO FINISH = 2 MEN/1 DAY	16	HRS.	$ 10 —		160 —	
LABOR TO STRIP FORMS = 1 MAN/½ DAY	4	HRS.	$ 10 —		40 —	
LABOR TO FINISH GRADE	8	HRS.	$ 10 —		80 —	
TOTALS:				3,168 —	1,080 —	60 —
THESE HOURS ARE FICTITIOUS						

Concrete Flatwork Estimating Sheet

PROJECT

LOCATION

DATE

BID ITEM	QUANTITY	UNIT	PRICE	MATERIAL	LABOR	EQUIPMENT
STORM DRAINAGE						
36" R.C.P.	220	Ft.	20 —	4400 —		
24" R.C.P.	100	Ft.	12 —	1200 —		
18" R.C.P.	300	Ft.	8 —	2400 —		
RAM-NECKS (PER QUOTE)	—	L.S.	—	330 —		
BACK-HOE 660 Yds ÷ 20 Yds/hr	33	Hrs.	10/20		330 —	660 —
LABOR TO INSTALL 4 Men	248	Hrs	10 —		2480 —	
BACK-HOE TO INSTALL	62	Hrs	10/20		620 —	1240 —
SPECIAL BACKFILL	130	Yds.	11 —	1430 —		
LOADER FOR BACKFILL	62	Hrs	10/40		620 —	2480 —
TAMPER	62	Hrs	10/10		620 —	620 —
CATCH BASINS ("A")	3	EA	450 —	1350 —		
INLETS.	1	"	1250 —	1250 —		
· BACK-HOE	12	Hr	10/10		120 —	120 —
LABOR TO INSTALL	32	Hrs	10 —		320 —	
" TO BACKFILL	16	"	10 —		160 —	
TAMPER.	16	"	10/10		160 —	160 —
MISCELLANEOUS	—	L.S.	—	150 —		
TOTALS				12360 —	5430 —	5280 —
THESE HOURS ARE FICTITIOUS						

Storm Drainage Estimating Sheet

How to Do a Unit Price Bid

The other type of bid which can be requested is a unit price bid. This kind of proposal is my favorite for several reasons. First, it is less risky because the contractor is bidding according to his takeoffs. When they do the work, they are actually paid for what they install. If they install more, they get paid more, and vice versa. It is also less time consuming since there is less quantity takeoff work to be done. Secondly, I like it because contractors can stack the bid in such a way the job can be their banker. Thirdly, a contractor can negotiate a unit price bid more easily and advantageously. Those things will be discussed in a moment.

For a unit price bid, use a thirteen-column accountant's sheet. For the sake of space, it has been compressed into a smaller sheet as shown in "The Sample Bid." This sheet is a recap spread sheet on which one can do his final work a few hours before the bid.

Some people do not use such a sheet. They just do the work for each bid item on a separate sheet, then mark it up on the sheet and arrive at a unit price for that item. They then just fill in the bid items and run a total on the proposal form. I don't like this method for a couple of reasons. First, because when they are all done with the bid, they do not know what their total material, labor, equipment and subcontractor ratio is without totaling the figures off of each separate sheet. They also do not know their total overhead recovery and profit without totaling the figures off of each separate sheet. When a contractor is done with a bid, and just before he decides profits and or contingencies, those are important things for them to know.

Also, this recap sheet gives a contractor the ability to easily mark it up for negotiation and so they can use the job as their banker. In addition, the recap sheet is more mathematically accurate and it allows them to figure their general conditions as one item and include it in their markup over the entire job.

If you will look at the illustration, you will notice the left-hand side is copied right off of the proposal form with the breakdown of items which they are requesting prices on using their quantities. To the right of those numbers are columns for the bid figures for material, labor, equipment and subcontractors.

Once an estimator has completed the recap sheet, he again make individual worksheets for those items for which he will be supplying material, labor or equipment. He then numbers them and does a very important thing: staples them together.

After he has completed all worksheets and totaled them, they can transfer those material, labor, equipment and subcontractor totals to the recap sheets. Then, he can total each bid item from left to right to arrive at a total cost per bid item. He can also total all the material, labor equipment, and subcontractors down each column to get a feel for the ratio between each. He can then total those totals across and total the total cost column down. The result must equal the total of their totals for material, labor and equipment (or they have made a mistake). As long as those numbers match, they know that the chance of their having a mathematical error in their recap sheet is zero to none.

They then begin to add on to the totals on the bottom of the sheet. The material, labor, equipment and subcontract costs of general conditions are added on to the totals. Then the sales tax (if any is required), labor burden, and overhead recovery figures plus profit are added. This will give them their total bid and an immediate feel for what this job will recover in overhead and have as profit.

Now, it is time to spread the mark up over the costs so one can come up with a total bid price per item. For the sake of example, let's say that the total direct job costs are $1,000,000 and the bid is $1,300,000. One must then spread $300,000 over the $1,000,000 in costs. "Easy," you say, "Just add 30 percent to each item's cost and that's your mark up" Wrong. Here is where a contractor wants to mark it up with two purposes in mind.

The first purpose is to use the job as a bank. When the job is 10 percent done, they want 25 percent of their overhead and profit. When it's 30 percent done, they want 50 per cent, when it's 50 percent done, they want 75 percent and when it's 75 percent done, they want 90 percent of their overhead and profit.

Let's say that on the above illustration the mobilization costs are $10,000. One would mark that up $10,000, and then mobilize in three days. If clear and grub costs were $10,000, he may add $10,000 to the clear and grub and do this item in one week. If earthwork costs were $1,000.000, he might add $70,000, and do the earthwork in six weeks. Then on those items which were near the end of the job he may only add 10 percent.

When a contractor does this, it is important he check the quantities of those items he is going to load heavy or light. They don't want to add extra money to an item which they have a bust on in their takeoff. For example, if the earthwork, by their takeoff, showed there was only going to be around 70,000 yard of dirt to be actually moved, any extra money which was put on the earthwork per yard would be lost on the 30,000 yards shortfall. In fact, if that were the case, he would put a little less than normal profit on the dirt and make some extra money on the job by putting this money in an item which will end up being the bid quantity (or will even over run). The reverse of this scenario would also hold true. If there were any items, even if they were at the end of the job, which were obviously going to be more than bid quantity, they would add a normal profit to them. This would not affect their total bid but, when the job was over, it would give them a higher profit. Now, what has just been explained to you is done with some risk involved. If the items are changed by change order, or things under run, even though the take off didn't indicate it, this method can cost the contractor some money. It is, at best, a calculated risk. However, it has paid off more than it has hurt.

The second thing which must be taken into consideration when marking up this job is doing it in such a way the contractor will be in the driver's seat for negotiations. For six months every job is bid is under the architect's/engineer's estimate. Then, for six years, everything which is bid is over the architect's/engineer's estimate.

Let's say, on my previous example the bid is $1,300,000 and the budget is $1,150.000. Do you know what a contractor is giving them when they give them their unit price bid? They are giving them a shopping list. They might as well give their kids a Sears catalog, their charge card and a budget of how much they can spend. And, that is just how the owner will look at the bid. They will look to see what they can afford from this bid and still build a project for their budget.

So, when a contractor is marking-up the job, he wants to anticipate what the owner will take out so he can put very little markup on those items and will put profit on the items which he anticipates the owner will keep. In order to do this, he has to put on an architect's/engineer's hat, or, put on an owner's hat. Then a contractor has to think like they think and ask himself, "What would I remove as an architect or engineer, or as an owner?" With this in mind, he should put less than normal mark-up on the items likely to be removed and put more mark-up on the items which must be installed.

Well, once a contractor has spread his markup over the items, taking into consideration the items discussed, he totals his markup column to make sure it is $300,000. Then he adds his markup on each item to the cost of each item to come up with a total bid price per item (see "The Sample Bid"). He then totals the column (cost with markup) down and it should total $1,300,000. If it does not, then there is a mathematical error in the figures. Then he can take each item total (cost with markup) and divide it by his quantities to come up with a unit price. If his unit price is 88.5 cents per foot, or $1.67 per yard, round it up.

Once these unit prices have been established, he can multiply them by quantities and come up with their price in the far right column. Then total the far right column down and it should total around $1,300,000. Again, if it does not, he has made a mathematical error. However, once it does, he knows the the chance for a mathematical error is zero to none.

Now he is ready to fill in the proposal, which will be the two far right-hand columns. But, let me point out something very important here. Filling in the proposal form from this recap sheet is not a one-person job. It is a two-person job. One person reads the item, then the unit price, then the total. The other person writes the figures in and repeats them out loud as they write them down. This is to prevent any errors in filling out the bid form.

The following page is a completed sample of a unit price bid.

ITEM	QUANTITY	MATERIALS	LABOR	EQUIPMENT	SUBS.	TOTAL COST	MARK-UP	TOTAL WITH MARK-UP	UNIT BID	TOTAL BID
MOBILIZATION	L.S.	-0-	4,000-	6,000-	-0-	10,000-	10,000-	20,000-	→	20,000-
CLEARING	10 ACS	-0-	3,000-	7,000-	-0-	10,000-	10,000-	20,000-	2,000-	20,000-
EARTHWORK	100,000 YDS	5,000-	35,000-	60,000-	-0-	100,000-	60,000-	160,000-	1.60	160,000-
CONC. STRUCTURAL	200 C.Y.									
CONC. FLATWORK	6,000 SQ. FT.									
RESTROOM BUILDING	L.S.									
18" R.C.P.	200 FT									
24" R.C.P.	260 FT									
30" R.C.P.	420 FT									
STORM INLET	5 EA.									
MANHOLES	3 EA.									
4" P.V.C. MAINLINE	2,100 FT									
3" P.V.C. MAINLINE	1,800 FT									
2" P.V.C. LATERALS	3,100 FT									
1½" P.V.C. LATERALS	5,400 FT									
1" P.V.C. LATERALS	9,200 FT									
ROTARY HEADS	120 EA.									
2" VALVES	12 EA.									
CONTROLLERS	1 EA.									
3" TREES	40 EA.									
2" TREES	90 EA.									
SHRUBS	260 EA.									
SEEDING	10 ACS									
TOTALS		500,000-	250,000-	150,000-	100,000-	1,000,000-	300,000-	1,300,000-		1,300,000-
GENERAL CONDITIONS		5,000-	20,000-	20,000-	-0-					
GRAND TOTAL		505,000-	270,000-	170,000-	109,000-					

· TAX · OVERHEAD   · L.B. · OVERHEAD   · OVERHEAD   · OVERHEAD   PROFIT/CONTINGENCY = BID ▼ 1,300,000 -

TOTAL ·   TOTAL ·   TOTAL ·

Recap Sheet For Unit Price Bid

Deciding What to Bid

How do you decide what projects to bid? There are three things to be considered. First, does the job fit your company and give you an advantage? Second, do you have the time to do the job? Third, does the job fit your work flow chart? The first two principles I will cover in this section, and the third in the following section.

Does it fit your company and give you an advantage? I have a definition for estimating which is fundamental to our understanding of this concept: Estimating is the science of arriving at what a job *costs your company*.

It is important we understand the difference between estimating and bidding. Estimating is taking a set of plans and specifications, and arriving at what a job costs in the big four: material, labor, equipment and subcontractors. Bidding is taking this estimate and adding overhead, profit and contingencies to it and turning in a price — a bid — for what you want your company to be paid for the job.

You can lower your bid price on a job by asking for less profit or by reducing your company overhead. However, you can never, nor should you ever, lower your estimated costs. Costs are costs. Therefore, estimating is the science of arriving, first and foremost, at what a job costs. From those costs, a bid is developed.

Several hundred years ago, the architects of the world may have gathered in a back room and conspired against all the contractors of history. The conspiracy was "competitive bidding," the process we're talking about.

You and I have heard these words used for years; words which naturally make us think we are in some sort of athletic competition.

Imagine we all get into a wrestling ring and strip down to our waists. Then we lie, cheat, bite, pinch, and gouge at the bid and at each other until only one of us, all bloodied and beat up, is left standing: This one is the low bidder. He wins the competition and as a reward, gets to go out and work hard and lose money on a job.

If you fall for the concept that bidding and estimating are the same thing—it is one company competing against another—then I can tell you what may easily happen to you. You will want a certain job very much. You will want to beat the competition, so you will begin to lie to yourself as to the level of costs. You will say things will cost less or you will buy less than they will or you will. In order to be low bidder, you will do this and, many times, you will *be* low bidder. However, when all is said and done, the job will cost you what it actually costs, regardless of what you said to yourself while you were estimating. Costs are costs are costs are costs! Estimating is the science of arriving, fairly and honestly, at what those costs are going to be.

Now, believing this to be so, if I were to give two different construction companies the same set of plans and specifications to estimate, they would come up with different costs for the same job. Why? Because one of them messed up? Not necessarily. They should come up with different costs because of the differences in their companies. One company, for example, will be able to do the job faster and cheaper because it has certain supply sources or possibly a crew of people who specialize in a certain work function or a piece of equipment which can get the work done cheaper than another type of equipment or method. This company, then, should be low bidder. Not because he underbid the other company, but because the job fits his company better than it fits anyone else; thus, it costs him less.

With this in mind, one of the first questions you should ask when you are deciding on what to bid is, "Does this job fit my company, and does it give me an advantage?" If it does, the job should be given a high priority on your bid schedule. If it doesn't, the job should be given low or no priority on your bid schedule. Thus, you will be spending your time on bidding the jobs where you have the greatest advantage and the greatest potential for being successful.

What are your company's advantages? Do you know what they are? If not, spend some time to determine them, and then look for them on the jobs which are available for you to bid. Play to your strengths!

One of the questions I am often asked is: "What do you do when there are 10 to 15 bids for one job?"

Increase your advantages! The construction industry, I believe, goes in several cycles. One cycle involves the amount of work available in relationship to the number of contractors who need that work. Of course, this is based on the economy and on the growth pattern of a given area. For instance, for several years there are often only two or three bidders on each job and things are going well. Everybody is busy and making money. Then, for a year or two—or even longer—there are many contractors at every bid opening. What do you do during such times to be a survivor?

There are three things to consider which can help you to survive. First, there is what I call people and crew potential. I know a contractor who started out as primarily a landscape contractor. But, he added an excavation superintendent and equipment to his company. Then he added a concrete superintendent and a building superintendent who could build small buildings. This group of people allowed him to become a general contractor on site-development-type projects. He was able to go out and bid projects with earthwork, concrete work, small buildings, landscaping and irrigation work and do all the work with his own people. This gave him a distinct advantage at the bid table. Not because he was cutting his throat in the areas of overhead and profit, but because he had an advantage at the level of cost.

What kinds of things are you doing with your own people? Are there some types of work which you could do on some of your jobs with your people rather than subcontracting them? If so, will doing that give you a decided advantage at the bid table?

A second thing which can give you an advantage and help you survive is to consider traveling to better work locations. I remember when a new company started up in a city and I was working for another company in that city. The new company took everything in a particular line of work which came out to bid for six months. We could not get any jobs. We said that if we needed their kind of work done at our office building, they could do it cheaper than we could at cost.

At this point, we had a choice. On the one hand, we could get down in the mud and take jobs at their prices and hope we could out last them. On the other hand, we could cut back on overhead and profit and look for work somewhere else. We decided on the second option. For a year we picked up and did jobs in five neighboring states. In nine months the new company which was picking up everything went broke and we came back home to a better market. We survived!

You see, I have a fundamental belief which is tied into my definition of estimating. When times are tough, you don't get down into the mud of cutthroat bidding. You can cut your profit as much as possible and reduce your overhead as much as is reasonable, and then you must look for advantages. You never, ever fool yourself and alter known costs to get down to someone else's bid. Costs are costs are costs, and they should never be changed. Estimating must always be the science of arriving at what a job really costs your company. What you do with it from there becomes the science of bidding; i.e., turning in a bid which you hope will get the job.

The third thing to consider is technology. I know a man who was always looking for new construction techniques and specialized equipment. When something new came out, he was there looking at it with only one purpose in mind, would this new technique or specialized equipment give his company an advantage at the bid table? If it did, he brought it home to his company. In his particular area, he was the first to use or implement new ideas and equipment which later became standard in the industry. Since he was one of the first, they gave him advantages for many years while others were still doing it the old way.

When deciding what to bid based on this thought, you should ask: "Does it fit our company and give us an advantage at the level of cost?" Next time you open a new set of plans and specifications, look them over with this very thought in mind.

The second thing to take into consideration in deciding what to bid is whether you have the time to do the job well. I helped a company bid a job which I feel typifies the need for this consideration. The job bid at eleven o'clock on a Monday morning in a small town two hours from this contractor's office. This had to be the worst time I know of for a bid opening! It was a large project, and by Friday afternoon our contacts of subcontractors and suppliers told us there were three other bidders besides ourselves who were calling around and getting prices. Three other bidders had spent time and money to do a very comprehensive takeoff and make all the necessary telephone calls.

At eleven o'clock on Monday morning, an estimator from this company laid down a sealed bid next to only one other bid. Two bids were all there was. Do you know what I wanted to do when I found out there was only one other bidder—besides raise our price? I wanted to jump in my car and go to the offices of the two contractors who didn't show up. And do you know what I would have found? I would have found thousands of dollars worth of work spent on the bid thrown in a trash can, and the reason might have been because they ran out of time. They got almost everything together, but eleven o'clock Monday morning rolled around and they didn't have it all together, and so they threw it away without turning in a bid.

I wonder how many thousands of dollars are wasted everyday in this country from uncompleted bids which get thrown away. Furthermore, I wonder how many contractors lose something even more valuable than a few thousand dollars, namely their reputations.

I remember doing a seminar in which I mentioned this very thought. I noticed an architect nodding his head so hard that I thought his head was going to come off. I sat with him at lunch. He leaned over to me and said "What you talked about this morning just happened to me yesterday. I have a very wealthy client in this city to whom I promised four bids for his project from the best contractors I could find. They all promised me a bid. Only two showed up with bids. The other two will never, ever see another one of my bids!"

I pride myself on very little in this industry, but on one thing I do pride myself. I cannot remember ever being late for a bid opening. I cannot remember one bid which I started on and did not finish due to lack of time. Let me give you some ideas I have implemented to help me accomplish that.

One of the first things I do when I receive a new job is to estimate the amount of time it will take me to bid the job. Will it take me five hours, 15 hours or 40 hours? Is a lot of the takeoff work already done, or do I need to do a lot of the takeoffs? Is the material to be used on the job from my normal sources or do I need to make a lot of telephone calls to find prices? Will the job require my normal subcontractors, or will I need to find new ones? All of these things are a part of the consideration of how long it will take me to bid the job. Once I arrive at that estimate, I then check my calendar to see if I have the time—or if I am willing to commit the time—necessary to bid the job. When I approach a job with this predetermined knowledge and commitment, I find I can do a better job.

One of the most important attributes of an estimator is organization. This is especially true when it comes to his time. I have always lived strictly by a calendar which, for the most part, is planned days and even weeks in advance. I have always used a calendar with large pages given to each day. Please turn to the illustration in the back of this section (page 514).

Notice that three times have been set aside for what I call garbage in, garbage out. Those are times spent in answering superintendents' questions, in receiving or making telephone calls, reading mail, and a myriad of other items. These times are placed through out the day at strategic times; i.e., when the day begins, at the end of the morning, and before going home at night. Two major blocks of time are reserved for estimating in the middle of the day. During those times you should be considered unavailable, except in cases of emergency. Three hours of uninterrupted estimating time is worth a whole day which is filled with interruptions. Plus, if you are working on a takeoff or a labor-pricing strategy and leave to answer a telephone call, you run a great risk of making a mistake. Notice the block of estimating time titled "evening hours." This time should be used for catch-up or for that extra commitment it may take to get a job bid.

With a calendar like this you will be able to look at the time when a job bids and then go backwards and see if you have enough hours to estimate the job. If you do, fill in those blocks with the name of the job and then *discipline yourself* to use this time toward the job. However, if you don't have the time, then don't start the estimate. It is always better to bid a few jobs right than a lot of jobs wrong.

I also find it is important to check on your progress as you are going along. Do you know when most errors are made in bidding? They are made during the last hour before a bid must be turned in to the owner. Telephones are ringing, prices are coming in, important numbers are found to be missing and in need of figuring, lots of multiplication and addition must be done, and the estimator is going as fast as possible, hoping to finish on time. I have seen this very thing occur with my own eyes. I have walked into offices where I have seen estimators' eyes as big as silver dollars. Sweat is pouring off their foreheads and their hands are working the pencils and calculators so fast that smoke is rising from the paper and the machine. They are making mistake after mistake in the final hour before bid time.

To prevent this problem I find it is important to check on my progress as I am going along. I always find myself asking this question: "How much work remains to be done and do I have enough time between now and bid time?" If I do not, then I need to put in the extra amount of time necessary to complete the bid on time. I ask myself those questions long before the final hour before bid time. This way I can keep from being caught in a hectic and risky situation during the final hour.

I remember helping finalize a bid on a large project for a contractor some 30 hours before bid time. The people who had been working on the bid had simply started into it without any plan or any idea of how long it would take. It was a unit price bid with over 150 unit price items. The contractor met me at the airport and took me to look at the job site. He told me if I didn't feel like there was enough time to bid it, to tell him so and we would all go back home. I wondered why he said this to me. When we got to the hotel rooms he had rented, I immediately saw why he had said this. Bid sheets, plans, and specifications were everywhere. Nobody knew how much had been done or how much needed to be done. Their eyes were already as big as silver dollars and their hands were flying across the paper, making mistake after mistake.

The first thing we did was to reorganize everything and to make a unit price recap sheet. I then sat down and looked at what needed to be done and estimated how long it would take. I figured if all of us worked throughout the entire night we would be able to have it done by the next afternoon. We each took certain bids items and began to work on them. However, I did not leave it there. At nine o'clock, at midnight, at 3 A.M. and again at 6 A.M. I did rechecks to see if we were on schedule. This way we were able to work by a plan, with confidence, without anxiety and eventually got the job done.

An estimator's time must be disciplined and used as effectively as possible if he is going to do the best with his available time. In deciding if you are going to bid this job or another, estimate how long it will take and ask yourself, "Do I have the time?" If you do not have the time to bid it right and in a timely manner, do not bid it. In bidding, more is not always better!

Without question the most costly and wasteful use of your estimating time is misspent starting a bid you won't have time to complete.

How do you avoid this situation? <u>Before</u> you begin preparing any bid, ask yourself, **"Do I have the time to bid this job?"** If the answer is no . . . don't start! Here are some tips for you to consider:

1. First: estimate the total time it will take to prepare the bid from beginning to end.

2. Next: distribute in your Estimator's Calendar the hours you'll need to complete the bid. Monitor your progress and make time allocation changes if needed.

3. Set aside specific times each day strictly for estimating. Let others in your firm know you are not to be disturbed during this time period, unless there is an

emergency. By adopting this "closed door" policy, you'll be 30-50% more productive than if you permit interruptions throughout the day.

4. Do your "telephone" work (returning calls, answering messages, etc.) at set times. I suggest you do this in the morning before you close your door, and again later in the afternoon before close of business.

5. Use the evening hours as a "safety valve" for those jobs which take more time to bid than you estimated. With practice, your time estimates will become increasingly accurate and your "safety valve" session less frequent.

The Estimator's Calendar

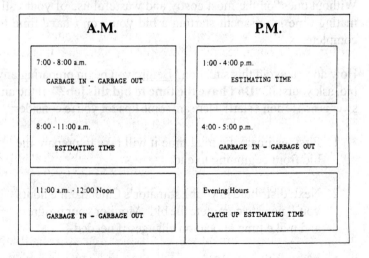

A.M.	P.M.
7:00 - 8:00 a.m. GARBAGE IN – GARBAGE OUT	1:00 - 4:00 p.m. ESTIMATING TIME
8:00 - 11:00 a.m. ESTIMATING TIME	4:00 - 5:00 p.m. GARBAGE IN – GARBAGE OUT
11:00 a.m. - 12:00 Noon GARBAGE IN – GARBAGE OUT	Evening Hours CATCH UP ESTIMATING TIME

The Work Flow Chart

The third thing to consider in deciding what to bid is the work flow chart. Because it is such a diverse tool, I have chosen to dedicate a separate section to the subject. However, before we do, I must rededicate the discussion with a couple of philosophies of mine.

First, I believe an estimator—or contract administrator as I prefer to call him or her—should employ as many tools as possible to keep involved in the mainstream of what's happening in a company. I think one of the worst things which can happen is to have the person who does the estimating stuck off in some ivory tower of a back office. Into one door march the plans and specifications to be bid. The contract administrator works up a bid and sends it out another door. All this person does for the company is put out numbers day in and day out. Those numbers will become as bogus as a three-dollar bill very quickly. Why? Because this person will lose touch with reality. They will not hear and feel and experience the harsh realities and difficulties of what it takes to get something built.

I have an education which is worth at least $200,000. I did not pay for it. The contractors I worked for did, God bless them. It is an education of mistakes, misjudgments, misunderstandings, and misapproaches. However, if I would not have learned from those mistakes, this education would be of no use to me or to anyone else. How important it has been for me to be a part of every problem so I could learn from those problems.

Anyone who does estimating needs to be aware of everything which goes on in a company. They need to know of problems when they arise and be a part of solving them. They need to know when and why jobs don't go as fast or bill as much (or vice versa), as was originally thought. Any tool which can help in this is a good tool, and that is what I believe a work flow chart to be.

There is a second philosophy I have which compliments the work flow chart. We in the construction field believe we are in the business of building things. We think our business is a product- and material-oriented business. After all, we take all kinds of different things and assemble them all together into one thing; e.g., a building, bridge, road, or park. Not so!! The greatest risk, the greatest problem, and the greatest "make it or break it" factor is *labor*. Material does not get drunk over the weekend and then not show up for work on Monday. Tractors do not get married to trenchers and have back-hoe babies. Neither do they break up with their girlfriend nor get a divorce. People do!! Labor is the biggest risk in the construction business!!

Construction is a people business. Contractors who realize this and know how to motivate people and work with people are successful contractors. Now, one of the most important things about people is *attitude*. If you get a company full of people who generally have positive, aggressive, get-it-done attitudes, you will not have a cash register big enough to count the money they make you. However if you get a company full of people who work with negative, complacent, we-have-all-day attitudes, you will not have enough money to buy a cash register. Yes, construction is a people business, and your people's attitudes will make or break your construction firm.

Now, keeping all of this in mind, let's look at the Work Flow Chart. To set up a work flow chart, I use a twelve or thirteen column accountant sheet. Each column is assigned a month of the year across the top of the sheet. Right under each month I put in a figure, which is the goal for the particular month. That is how much we need to bill in order to recover our overhead and to make a profit.

Before we go any further, I need to explain this figure and how we arrive at it. I believe a construction company must sit down and pre-plan a budget for overhead. Then it needs to establish how much work must be done to *recover* this overhead. Notice the word I have emphasized: *recover*.

Overhead is something which goes on whether you do any work or not. The rent, telephone, salaries, etc., are there no matter how much work you do. Consequently, look at overhead as a fixed cost, for the most part, which must be recovered if your company is to survive. Do not charge for overhead when you bid a job. Rather, if you get and do the job, recover a part of your overhead from this job. You establish an overhead figure for the company and decide how much work you can do with this overhead. Then, set up goals as to how much work you need to do in order to re-cover your overhead.

Let's say you feel you can do $1,200,000 in gross sales in a year. In another section we will discuss how you establish your over-head recovery numbers for this volume. But, for now, we want to spread $1,200,000 over a year's time. If you are a seasonal busi-ness, you distribute it according to your seasons and when you feel you will be doing more or less work. These figures become your goals for each month, enabling you to have a plan for your work so you can work your plan. If you are not meeting your goals, your overhead must be adjusted downward.

Going back to the work flow chart, you will notice on the left-hand side are all the jobs your company has under contract, with the dollar amount in thousands by the name of the projects. You then sit down with whomever is responsible for getting the work done in the field. You decide how each job is going to be done and what work will be done in each month. The contract admin-istrator, from his original bid, puts a dollar value on the work and then sets up a goal of when the job will bill and for what amounts. After this is done on all the jobs you have, each month is totaled and the contract administrator is then able to see what months met the goal and need no more work. They can also determine which months fall short and which months need more work for each time period. Thus, it becomes a tool to help in deciding what to bid.

It also becomes a tool to motivate and to set a good attitude in your field people. Remember, it was the person in charge of your field operations who told you when and what he would get finished on each job. That should become their goal and responsibility to accomplish; they should then be aware of what is expected of them. I used to make copies of this chart and make sure the field people had one. Then when the monthly billing was finished, we would fill in the line that says "actual." After we saw what the actual billing was, the next month's meeting over this chart had a discussion in it about how or why it was more or less than anticipated. Each job was then readjusted for each month based on the new plan of approach and billing for a job. The work flow chart becomes a great tool to gather people around to discuss what is really happening on each job.

It can also go one step further and be yet another tool. It can become an accountant's tool to inform the contractor, his bank, and his bonding company about how he is doing and will be doing financially. Two of the contractor's best friends in this business are his bank and his bonding company. Any tool you can use to show them you are a "front door contractor" and you know what you are doing, is a tool you should use.

Bonding companies and banks deal with contractors from two aspects. One is liquidity and the other is credibility. I find almost every contractor understands liquidity. It simply has to do with determining how much cash would be left if you were sold out in a day. But I find a lot of contractors do not fully understand credibility and how important it is to bonding companies and banks. If you know what is going on with your jobs on a month-to-month basis, and you are quick to know when a job is going bad, then you are a credible contractor. If you know what months you are going to make money and what months you are going to lose money, you are a credible contractor. If you understand your cash flow and when you will have extra cash and when you will need to borrow extra cash, you are a credible contractor. Bonding companies and banks will go alot further with bonding limits and loans to credible contractors than to those who do not know what is happening to them or who do not have the tools to perceive what is happening to them. Can you blame them?

Many contractors remind me of a guy sitting on a chair in a dark room. Someone or something knocks them off the chair, so they get up off the floor and get right back into the chair. Someone or something knocks them off again and once again they get up off the floor and get back into the chair. Many of them keep this cycle up until they go broke. The sad thing is if they had turned on a light they would have been able to do something about the someone or something which kept knocking them off the chair. A work flow chart and a cash flow projection are tools which will turn on the light.

In the second illustration in the back of this section (page 522), I have taken the same work flow chart and added a cash flow projection to the bottom. The first additional line of information requires the contract administrator to go back through his/her bids and come up with the direct job costs in the big four—material, labor, equipment and subcontractors—for each month. This cost is based on the anticipated billing. On the next line, insert the anticipated overhead for the month. On the next line down, insert the anticipated retentions on the billing. These three figures are subtracted from the billing. To this figure you add any retentions you hope to receive this month. The total on the bottom will tell you the amount of cash you can expect to have in the bank after you collect all your billings and pay all associated costs. This figure should not be confused with profit, because it has retentions and potential overbillings in it which will greatly affect it. This figure will only indicate when you should have money in the bank and when you can anticipate having to borrow money.

I will conclude this section with one final observation about these tools. Tools are for people, not people for tools. In other words, whatever tools you use should be tools you understand and like. No one should be forced to use something they do not understand—or like—because the tool will just become a millstone around his or her neck and eventually he or she will find enough reasons or excuses not to use it.

Today you can easily buy a large magnetic board for construction planning. I have been in some contractors' offices who use them with great success. Their people like them, understand them, and use them. I have been in other contractors' offices where these boards are merely expensive decorations used to write messages on or pin up notes. Why? Because their people tried and tried to use it, but it just wasn't them, so they devised other tools. None of the tools and forms in this book should be forced upon you or your people as they are. That would be making people for tools. Rather, you should take these tools and work them over until your people understand and like them. Then they will be tools for your people.

WORK FLOW CHART

GOALS	100,000	80,000	80,000	100,000	120,000	120,000	100,000
MONTH	AUG.	SEPT.	OCT.	NOV.	DEC.	JAN.	FEB.
ABC JOB (120M)	20,000	30,000	40,000	30,000			
XYZ JOB (50M)	10,000	30,000	10,000				
MNO JOB (80M)				40,000	40,000		
PQR JOB (50M)	40,000	10,000					
FGH JOB (90M)	30,000		30,000		30,000		
TOTALS	100,000	70,000	80,000	70,000	70,000	0	
ACTUAL	( )	( )	( )	( )			

Work Flow Chart Tracking Contracts to Goals

WORK FLOW CHART

GOALS	100,000	80,000	80,000	100,000	120,000	120,000	100,000
MONTH	AUG.	SEPT.	OCT.	NOV.	DEC.	JAN.	FEB.
ABC JOB (120M)	20,000	30,000	40,000	30,000			
XYZ JOB (50M)	10,000	30,000	10,000				
MNO JOB (80M)				40,000	40,000		
PQR JOB (50M)	40,000	10,000					
FGH JOB (90M)	30,000		30,000		30,000		
TOTALS	100,000	70,000	80,000	70,000	70,000	0	
ACTUAL	( )	( )	( )	( )			
COST	60,000	50,000	50,000	45,000			
PROFIT	40,000	20,000	30,000	25,000			
MINUS OVERHEAD	20,000	20,000	20,000	20,000			
MINUS RETENTION	10,000	7,000	8,000	7,000			
SUB-TOTAL	10,000	< 7,000 >	2,000	< 2,000 >			
RETENTION REC	— 0 —	— 0 —	10,000	5,000			
CASH	10,000	< 7,000 >	12,000	3,000			

Work Flow Chart Tracking Contracts to Goals, Including Cash Flow Projections

Minimizing Errors

There are three ways to minimize the errors on your bid: the first one is finding mistakes as quickly as possible, so a contractor can minimize the effect on the company.

When a contractor is low bidder, he should take on a certain attitude. Even if they are low by $35.50, they take on this attitude. It's an attitude that says, there's only one reason we are low bidder, we made a mistake. And after everybody shakes their hand and congratulates them, they are driving back to their office saying "Woe is me, I made a mistake."

They go right into their office and start looking over their bid so they can find their mistake. Don't ever file away the low bid saying "I know there's mistakes in the bid, we'll find them when they come around." You sure will! They'll show themselves at ten o'clock some morning and by two o'clock the contractor will be writing a check to pay for the problem; a problem they might have been able to minimize had they known of it earlier. The more quickly they find their mistakes—and there are mistakes in every bid—the better chance they will have to minimize the impact of those mistakes on their company by getting people involved in brainstorming ways to get around the problems.

There are two kinds of jobs in people's minds. There are those fat, juicy, good jobs which were bid right and have lots of money in them. Then there are the problem jobs. They're jobs which stink from day one and everybody knows it. However, time and time again, those good, juicy, fat jobs become losers. Conversely, those stinky, old, smelly jobs become money makers. Do you know why? Attitudes. People were concerned about this bad job. They babied it and nurtured it and finagled on it. They worked out a way to turn it around. But, the fat, juicy jobs, nobody worried about. They just waded into them with very little concern about costs, because they were so fat they couldn't lose. But, with this kind of attitude, they do.

There's a second important way to minimize errors on a bid. It involves the mathematical calculations. How many of you are reading this book have a doctorate degree in the calculator? No one does. It doesn't matter how dexterous someone is, they are going to make mistakes on the calculator.

One of the things about these estimating systems is where an hour or two before bid time an estimator is working on only one or two sheets. During that time, he can take his worksheets to someone else in the company and say, "Do me a favor, check all my math." Now let's say he pays someone $50 an hour to do this and it takes her 30 minutes to check his math. She will find a $25 or $2,500 or $25,000 mistake. When she does—and she will—he will have made the wisest investment of his life.

Every one of us has tunnel vision. We see a project as though we have no peripheral vision. We see very plainly what is straight ahead before us, but we cannot, nor will we ever, see it completely as it really exists. There will always be things around the tunnel we are looking past, and those things will affect and influence the cost of each job.

It doesn't matter how long an estimator looks at a job, whether it's six hours, six days or six weeks, he will never catch some of those things. He may have estimated a certain kind of material which is of questionable use, or may have figured on using a certain procedure which won't work. Maybe he planned on doing something with a certain piece of equipment which will not operate under conditions of that job. This is why it is important to take an estimate to someone and say, "Make me prove to you, make me convince you, it can be done for this price." And, if you can't convince them, then together arrive at what kind of cost should be used.

This concept really applies to the labor figure on an estimate. What's the biggest risk in construction? Labor. Because of this fact, it should be the most scrutinized part of a bid.

The best people to talk over a job with are field people. They will always see things an estimator can't see, so they are the best people to go over an estimate. Once they are convinced it can be done for certain costs, then it rests on them to go out and do the job for those costs. You see, I don't care how pretty and nicely an estimator does an estimate. If the field doesn't get it done for what was figured, it's not worth the paper it's written on.

Some estimators may feel this last procedure puts them in a subservient position to others. Hey, I'm the first to admit that no matter how good an estimator is, no one is perfect. We need to be saved from our own mistakes and let's not be too proud to put hundreds, if not thousands, back into our pockets by making ourselves justify our estimates.

Remember, estimating is a process! One of the steps in the process is to justify your costs in order to minimize errors.

You Are Low Bidder? So What!!!

One of the worst things you can do after you've turned in your bid on a project is to walk away from it, go back to your office, and wait for a contract. This is true even if you are the low bidder and there are no alternates up or down which would change this fact. If you walk away, you may never see a contract.

There are two major cycles in the construction industry. I have already told you about one. Now let me tell you about the second one. It seems for six months, everything you bid is under the engineer's estimate. Your bid is $50,000; they have a budget of $55,000. Your bid is $1,125,000; they have a budget of $1,200,000. Then for six *years*, everything you bid is over the budget by 5 to 20 percent.

I know when this happens. However, we can't seem to convince architects and engineers as to why it happens. Today they are designing projects which will not bid for a year, maybe two. They are using budgetary figures from cost indexes—or bids—which are six months to a year old. So, when the project comes out to bid, the budget is based on figures from two to three years earlier. No one can accurately compensate for the change in costs over this time frame. I recommend architects and engineers spend a few extra dollars and have an *estimator*—or an estimating consultant who is in the trade—put together their budget figures just prior to putting out a bid. This way, changes can be made before bid documents are assembled and before contractors spend money on the bid process.

Well, what should you do when you find out your price is the lowest bid? I always ask three questions. Question number one is, "What's the budget?" Even though a budget may have been published or spoken of earlier, I have found sometimes the actual construction budget is very different from what was published. Now, the answer to this first question is usually a figure which is lower than my low bid.

That leads me to my second question, "What are you going to do about it?" I also know the usual answer to this question. "We don't know. We had a budget of $_____. We'll have to take another look at it. We can only spend $_____."

Then I ask a third question, which concludes with an emphatic statement: "Can we talk about it? I will do you a job for $_____ (whatever their budget is)."

You must understand the psychology involved at a bid opening where you are low bidder but you are over the budget. At this moment, you are the only sane person in the room. Everyone else has the lights on, but nobody is home. It is up to you to pick up the fumbled football and run with it.

To further understand the psychology, we need to look at the frame of mind of the others. First, the architect. When your low bid is over his budget, one thought quickly flashes into his mind. You are a crook. You are trying to get rich on his project. It doesn't matter there are six other bidders with higher prices than yours, you are a crook. I've had them get up and look out the window to see if I drove up in a Mercedes 380 SL.

Secondly, we need to consider the owner. To do that you must imagine the following scenario. Let's say this someone from home called you at the office and gave you a list of five things you were to pick up from the store on your way home. If you didn't pick them up, no dinner tonight. After you get off work, you go by the store, pick out the items, and go to the fast lane to be checked out. The articles are rung up for a total cost of $13. You look into your billfold and there sits a lonely $10 bill, $13 on the cash register, $10 in your billfold. Ten other guys are all standing in line anxiously looking at you and they are wondering what the hold up is all about.

That's exactly how an owner feels at the moment your low bid is opened. $130,000 low bid, $100,000 in the bank as a loan commitment, $1,300,000 low bid but only a $1,000,000 budget. The owner has made commitments to his bank. He has done market surveys on what he can rent or sell his project for based on the local market. Everything revolves around this budget and now you want more money. Pick up the football and run with it by asking the three questions.

I remember bidding a job in March several years ago. The project was for the site development around some old houses on an abandoned street. In this particular city there were several satellite campuses from nearby universities. It was decided one major campus would be built near the downtown section of this city and then they would pool all of these satellite schools onto one campus.

Now, in this city there was a committee of high-society ladies who went about preserving old historical buildings for posterity. And, right in the middle of this campus project were eight old houses and an old neighborhood store. This committee of ladies pleaded with the owner not to tear down those old buildings. They offered to raise money to restore the buildings and build a park around the outside. They would then give the buildings back to the owner to be used as offices for the college. Great Idea! They established an August 1 date for a great ceremony in which they would hand back these buildings and park to the college. A big, high-society gala event was planned with important people from everywhere being invited. It was the middle of March and they were just letting the bid on the exterior site work.

I showed up with my bid ten minutes before the 2 P.M. opening. At 2 o'clock I realized I was the only dummy there.

"Mr. Vander Kooi, do you mind if we wait ten minutes to see if anyone else shows up?" they asked.

"Please do," I said. "In fact, I'll call some people if you want me to." (I don't like to be the only bidder.) Fifteen minutes went by and no one else showed up. It was decided they would open my bid anyway. When my bid of $225,000.00 was opened, I knew something was very wrong. The architect gave me a look I will never forget. It said, 'You're a big, huge crook!' I've never seen a look like this before or since that day. It nailed me to the wall and sent a cold chill up and down my spine. The chairwoman of the committee was also at the bid opening. She was wearing a mink stole and smoking a cigarette through a long cigarette holder. When they opened my bid she let out such a large sigh I thought the cigarette she was smoking would go flying through the air at me. I said to myself, 'Something is wrong. These people do not look very happy.' So, I asked my first question almost out of curiosity. "What's the budget?"

The high-society lady sheepishly responded, "$135,000.00." Ninety thousand dollars difference between my bid and their budget.

Trying not to be sarcastic I asked my second question, "What are you going to do about it?"

The architect could not even talk, so the high-society lady responded, "We don't know. We have raised money and we have commitments to our donors. We have a gala event all planned for August 1."

Now, I should have just closed up my briefcase and gone home. However, I talk faster than I think, so I had already asked, "Can we talk about it? I will do a job for you for $135,000."

It was Thursday afternoon, so we agreed to meet the next day at 9:00 A.M. in the architect's office. I came into his office to meet with him and the high-society lady at 9:00 A.M. sharp. The architect must have set his face in concrete all night because the minute I walked in I received the same look which said, "You're a crook!"

Well, I said, "Let's get started. We've got $90,000 to cut out of this job."

For the first time since the bid opening I heard the architect talk, as he said, "I've been thinking about this all night." 'I bet you have,' I thought. "There are three trees over here. I think we can get by with two. How much will that save us?"

"Two hundred dollars," I responded.

"Well, I've noticed some of these shrubs will really grow tall and I have them four foot on center. I think I can spread them out to six foot on center. How much will that save us?"

"Two hundred and eighty dollars," I said.

"These back patios," he said, "They're 8 feet by 8 feet. I think I can get by with 6 feet by 6 feet without hurting the integrity of the project. How much will that save us?"

"Four hundred and fifty dollars," I said. Two-and-a-half hours went by and we had only managed to cut $20,000 from the bid. It was 11:30 A.M. and my stomach was starting to growl for lunch. "Any more ideas?" I asked the architect.

Now, you have to understand when you start cutting on a project, you are actually cutting open the architect's chest and ripping out his or her heart. The architect has designed this project to look just right, just as it is. When you take things out, you are ruining his or her beautiful project and it is losing its integrity. You're slitting their wrists and they are bleeding to death.

"No," he said, as he sagged into his chair and I could tell he had been bled to death. My moment had arrived! You see, I had been thinking about it all night, too. I had been in my office early in the morning to prepare my own list of cuts. So, I reached into my briefcase to get my list with copies for everyone.

I said, "Here is a list of $90,000 worth of gutting (not cutting) you must do if you're going to do this project. I'm leaving for lunch. You think about it over the weekend. I'll be in my office Monday morning at 8 A.M. If you want us to do the project, you call me then and we'll have a crew out here on Tuesday and have it done by August 1. If you don't, re-bid the project; but I won't even be here next time." Then I got up and left.

At 8 A.M. Monday morning my phone was ringing. It was the chairwoman of the high-society committee and she was excited.

"Mr. Vander Kooi," she said, "We went out this weekend and raised another $10,000 and we want to award you a contract and buy back $10,000 worth of your cuts." Now we were starting a job on a positive note by adding back on to the job. We signed a contract, had a crew on the job Tuesday and finished two days before the August 1 event.

I'm proud of this job for two reasons. First, because it's a good-looking job. It's still there and I point to it with pride. Secondly, because it was the highest profit percentage on any job this contractor ever did. In fact, the profit exceeded the cost.

It was only possible because I asked three questions. What's the budget? What are you going to do about it? Can we talk? I will do you a project for *whatever the budget*.

Getting Your Fair Share

I often find myself in a conversation with a contractor who is concerned he is not getting his fair share of the work he is bidding. This can often be due to a misperception of success by the contractor or possibly a misunderstanding of the construction climate in which he is uniquely involved in his area. Let me share with you the reality, as I see it, around the country.

First, just because you are not getting as many of the jobs you are bidding as you think you should, does not necessarily mean you should lower your prices. In fact, most of the time, the better solution is to bid more work, thus raising the amount of work you are successful on, simply by increased volume. I find, in most cases, it is cheaper to bid more than to not make as much money on what you might sign up at lower prices.

Second, as a rule of thumb, I tell contractors in a competitive bid market, if you are successful more than 15 percent of the time, you could ask for more money. If you are in the kind of specialized market where there are very few contractors in your area or you design and build and sell your work after developing a trusting relationship with your clients, this percentage may get as high as 40 to 70 percent.

Third, you may have some flaws in your estimating strategy which need to be corrected. There are four places in the estimating system on page 434 you need to take a good hard look at in order to see if you are having a problem in this area. First, you may be either overly optimistic or pessimistic on the number of production hours it will take to produce the product. This system is heavily dependent on estimating the labor properly. Any deflated or inflated figures at this beginning level becomes inflated by 100 percent by the time they come out in the final number. Second, your labor burden may be higher because you are paying your people too many benefits in comparison to what other contractors are paying their people for the same amount of work. Third, you may be paying your people more per hour than other companies do, to people who produce as much per hour.

I never condemn a contractor, however, for these last two items, because they can happen so easily. You see these flesh-and-blood people everyday. They come to you one by one with their needs, and naturally, you want to help them. But this help can sometimes cost you your ability to be competitive.

Fourth, your overhead may be too high or too low in relationship to the amount of work you intend to do, and in relationship to how other people are running their companies and the overhead they have. Fifth, you may be asking for more profit than the economic climate in your area will allow. Remember, the system on page 434 works. If it does not work in the marketplace, then these items need to be considered.

At the end of the day we opened the bids. Out of 16 contractors, guess who was low bidder. He was! Do you know why? It was because his people could out-produce everyone else's in the room! And the way he ran his company gave him the lowest overhead-per-labor-dollar I have ever seen. The system works! If you will be true to it, it will be true to you.

Using this system and strategy, you can get your fair share of profit even in bad economic times. I had a certain contractor come to a workshop several years ago. When I looked at his books I saw he had done almost $750,000 in sales, with a profit of only $7,000, which became his salary. We set him up on our system, and a year later he returned to tighten up his numbers.

He said to me, "When I got your brochure about the workshop, I thought about how you wanted 5 percent of my yearly salary to attend, but I did it anyway." He then pushed his recent financial statement across the table to me. That year he only did $650,000 in sales, but he was able to pay himself a decent salary and still make a 4 percent profit. He went on to say, "We are going into a tough economic time in this area and I am here to lower my average wage, my labor burden, and my overhead, so that I can survive." Six months ago, I talked with him on the phone, and in a tough economic environment, he did the same amount of work, paid himself the same salary, and increased his profit to 8 percent! You can get your fair share, at a profit, and receive a salary just like everyone who works for you does.

Fourth, many times I find contractors lose perspective on how much they have bid. Or else they do not follow up properly on their submitted bids. The everyday pressures and concerns of running their business soon distract them from bidding enough work or from adequately following up on their existing bids so they can supply their needs. The illustration on page 538 is of a bid board that I would recommend you establish. The bid board is designed to help you quantify and measure the dynamic process of obtaining new jobs. It helps you to set goals and to quickly identify where you need to focus management attention in this crucial area of your business.

Obtain a white board which can be erased with a cloth. These can be purchased along with special pens (two black and one red fine point) at almost any office supply store. Fill in the board as shown in the illustration. Using your black dry ink pens, print the letters in neat, block-style print about one inch high.

In Area One, enter any place where you might possibly obtain sets of plans to bid, or where you might obtain leads for getting new business. Plan rooms, dodge reports, green sheets, local government facilities, general contractors, developers, and new projects you see in the paper or in a neighborhood, are just a few places you could enter in this section. You should also put names and phone numbers of the ones which will need to be continually contacted.

In Area Two you enter the project name as soon as you are able to get a set of plans to either bid, or to design and bid. This helps you to monitor your bidding/designing backlog. Depending on your size, you should always be striving to have five to ten projects (at a minimum) in Area Two. Once you bid/design a project and submit it, erase it from Area Two and enter it into Area Three.

Area Three lists those completed bids/designs which have been submitted but have not yet had contracts awarded. Include your bid amount and any remarks you feel pertinent along with the project names and those you can contact concerning this particular job. Once the contract has been awarded (hopefully to you), delete it from Area Three. You should have room for 15 to 25 projects which are awaiting contracts. Total the amount of these projects.

I now want to explain to you how to determine your projects awaiting contracts goal. This goal shows you the total dollar amount of the projects awaiting contracts which you should have on your board at any given time. Let's use the following example. If your sales goal for the year is $500,000, and if you are awarded one out every ten bids you submit (which is 10 percent or .10 as a decimal), then divide $500,000 by .10. The resulting figure of $5,000,000 is approximately the amount of work that you will have to bid over the next 12 months to obtain $500,000 in sales. Next, estimate the length of time (on average) it takes, between the submission of your bid, and the awarding of the contract. If it is roughly 30 days, divide 365 (the number of days in a year) by 30 and you get 12.17. Round off the answer to 12 and then divide to get 416,667. Thus, at any one time, you should have a little over $400,000 in bids which have been submitted and are awaiting a contract to be awarded.

You do not have to be exact with these calculations, as they are not precise quantities. However, the formula can help establish a goal. For instance, if you arrived at the $400,000 figure for your company and you thought it was too low, then bump it up to

$450,000 or $475,000. Once you get a feel for the process, you can fine tune the numbers. Do not get lost in the calculations, or carve your goal in granite, for it is not necessarily a static "right" number just for you.

Once you have arrived at this goal, enter it on Area 3 under "Goal." Subtract this figure from the total dollars of the bids/designs awaiting a contract. If your goal is greater than the total dollars awaiting contract, enter the difference in red and then get out there and submit some more bids. If it's less, then you know you are bidding enough, and you need to focus on follow-up with the ones which are out there.

Finally, fill in your fiscal year sales goal. If your goal is to complete $500,000 in work, put this figure as your "FY-9_ sales goal." Keep a running tally of your contracts won for the year (jobs that will be installed in this fiscal year) and enter this figure on the "Contracts won" line. Subtract "Contract won" from your "FY-9_ sales goal" and enter that figure (in red) on the "Additional work needed" line. Adjust these figures every time you sign a contract which is to be done in this fiscal year.

Be sure to put the bid board in an area where your eyes will naturally see it during the course of every day. I suggest locating it at eye level within a few feet of your—or your estimator's—desk. If it is out of sight, it will be out of mind. However, if you use it correctly, it can motivate key people in your business as everyone learns what they must do to meet the goals which have been set for the company.

There are no secrets to getting your fair share of the work which is out there. If you are going to be successful at getting your fair share, it will require a combination of hard work (bidding enough jobs) and continually watching and fine tuning your system, strategy and numbers, based on what is happening in your marketplace.

POTENTIAL NEW BUSINESS	PROJECTS IN OFFICE	BIDS/DESIGNS COMPLETED AWAITING AWARD		
		PROJECT	REMARKS	AMOUNT
			TOTAL $_____	
			GOAL $_____	
			DIFFERENCE $_____	
			FY -8_ SALES GOAL $_____	
			CONTRACTS WON $_____	
			ADDITIONAL WORK NEEDED $_____	

Bid Board Example - Size would normally be 3' high x 4' wide.

Contracts and Change Orders

One of the most important parts of the construction business is the contract itself and the changes which occur to each contract as the job progresses. Many times, the financial viability of a construction company is not just determined by how well they perform the principles which have been laid out thus far, but by how well they write contracts and deal with the changes to the contract.

There are many different forms of contracts available in the marketplace. However, it is this author's opinion one of the best is the AIA Contract which is available through most stationers or through the American Institute of Architects. It is also very wise to have an attorney look at whatever contract is to be used, or help a contractor formulate one.

It should be noted there are four generally important parts to a construction contract. The first is the legal names and addresses of the parties who are entering into the contract. This should also include the legal address of the project if it is different than the owner's address.

Second is the work that is going to be done, and the agreed-upon price for that work. This section should either list out, in detail, all the items which are included in the price or refer to a set of plans and specifications that detail the work to be accomplished. Many a contractor has gotten into legal trouble with a client for not being very specific in this area.

Also, the exact price for this work should be laid out. There is always the question of how detailed should a contract get in showing all the different breakdowns of prices in their proposal and or contract. It is generally believed the less a contractor details a price, the better off they will be in the long run. The reasons for this are several. The more a contractor tells people about the break down of prices, the more ammunition they give someone to use against them in the long run. Also, when too many unit prices are given at the time of a bid, the owner has an opportunity to go shopping.

The best contract and proposal is one which does not give too great a detail on the breakdown of the price. Either it shows one price for the work or it is broken down into phases such as landscape, irrigation, hardscape or it is broken down into front yard and back yard work. This will give the owner enough information to make a decision. However, the contractor should have a general feel for what can be done to reduce the price and the amount of those reductions so they are able to negotiate the final price while meeting with the owner.

This approach often engenders argument from both contractors and owners who are used to or prefer unit prices. No one buys a new house and is told how much the sinks, doors, drywall and front steps cost. They buy a house based on the price of the house and very little of the break down of the price is offered by the builder. When a car is bought, no one asks how much the front fender, carburetor or back seat costs. They buy the car at the final negotiated price, period. Why should our contracts be any different?

The next part of a good contract should deal with the fact changes will be made and the method by which those changes will be handled from a price standpoint. People plan, but a fact of life is plans change. No matter how well the plans are done and no matter how much everyone agrees with the plan before work starts, changes will be made to those plans. A little later in this section, change orders will be discussed.

The best way to deal with this issue is to establish a method by which these change orders will be handled. One way is to stipulate they will be priced before work is done and no work will be done until this price is agreed upon by both parties. If this is done, a time limit for approval should be established so the job is not shut down for an extended period of time while the price is being settled.

The downside of this approach is owners can become very defensive of the price. Most owners do not trust a contractor and the distrust becomes even greater when a negotiated price for a change order is being discussed.

Another method to reduce this problem is that the contractor includes the prices to be used for change orders in the contract. A percentage to be added to the cost of materials is a part of the contract. An hourly figure for labor, including the cost of the truck, is a part of the contract. A price per hour for the use of equipment like Bobcats, trenchers, etc., is a part of the contract. An amount to be added to the changes in a subcontract is a part of the original contract. Then when changes occur, the material, labor, equipment, and subcontractor costs are tracked and these agreed upon percentages or prices are applied to the extra work. The owner will want to know, and should be supplied, a rough price for the change order. However, contractors should include in the price a statement which it is a rough price and the final billing will reflect the actual costs and time spent.

The last important issue in a contract is the method of payment. If a deposit is required, the amount should be stipulated. If the owner will be billed on a regular basis, the time frames of the billing and when payment is due should be stipulated. If payment is not received under those terms, the contractor has a contractual reason and responsibility to cease work until this condition is met by the owner.

Change orders are a part of the contracting business, yet they are a risky part of the business. At no time should a contractor enter into a contract believing he will make money on those change orders, and can therefore lower their price on the original bid. Owners and architects are very skeptical of change orders. Consequently, who is to say a contractor will get a change order or will get it at the needed price? Most lawsuits between contractors and owners today revolve around change orders. Many times contractors under price change orders or, if they price them properly, the owner is unwilling to pay those change orders.

Many change orders require the contractor to be aware of two things. First, be certain to visualize all the costs involved in the change. Usually, changes cost more than the original work because they inhibit the natural flow of the work as it was originally laid out. Also, changes may require the contractor to change other items which are not obviously connected to the change but will cost the contractor money. Before the price is given, a contractor must really consider the change will affect the job.

Second, a contractor should never, repeat, never, do work affected by an owner's request to change the job until he has a signed agreement the owner acknowledges the change and the change to the price. Many a contractor has lost money and never collected for change orders because this was not done. Owners may have given a verbal consent to the change and the price increase; however, when it comes time to submit and collect the bill, they can have a convenient loss of memory as to their approval. The payment of such change orders often becomes something the owner feels is a negotiable item before making final payment.

Just as the contract details the work to be accomplished, a good change order should be very detailed as to the changes which are to be made. These details should be accompanied by a drawing or sketches of the change, and those drawings or sketches should also be signed.

A good contract and properly signed and documented change orders will give a contractor a much better chance of making money and being paid upon the completion of a project!

True Life Stories From The Real World

✦ *We Went To Hell And Back*

It started out eight years ago as a Ma and Pa operation. Ron and Bonnie were happy-go-lucky people who wanted to start a business of their own. And what could possibly be more glorious than landscaping?

Unfortunately, they had problems before they started. It is called "lack of business." Neither of them knew much about business. After all, as Bonnie stated, "we were landscapers, not business people."

But they were young, energetic go-getters who were going to make their mark in the industry. It started out as small residentials, which became very short-lived. They wanted to grow fast and play with the big boys. So they got themselves into the commercial side of landscaping.

Not only was the business growing at this point (or so they thought), so was their family. They now had two babies (constantly crying as Bonnie stated), they were working 60-hour weeks plus a lot of additional night work, and to put the icing on the cake, decided to build their own custom home. Yes, ladies and gentlemen, welcome to the wealthy world of the landscape contractor!

After two years, they suddenly noticed the black was turning to gray, then to white, then to pink, and suddenly it was ALL red! They knew they had to do something fast, so they decided to attend our Estimating Workshop. At the end of the second day, they turned in their bid like everyone else and sat back and smiled. They had done their homework and knew they were going to be low bidder. Their prices were good (so they thought) and they would make a profit off of the job. Then all of the other bids were opened and suddenly a little bell went off in their heads. Why were their numbers so incredibly lower than everyone else's?

They went home that night and decided to refine some calculations and attempt to lower their overhead figures.

Things started to look better for Ron and Bonnie. Business was getting better and it appeared to them they were making money. But keep in mind, they still had no real "business sense" and were not sure how to track their progress. Then along came what they considered to be their golden opportunity. Now they were really going to get to play with the big boys.

A substantially large commercial project was coming to bid. It would come in at over $200,000 and it was going to be a landscaper's dream! The company would supply the labor and equipment and the plant material would be supplied for them. Oh yes, they were going to be in hog heaven! They sharpened their pencils, checked all the figures and produced their bid. They couldn't figure out why the general contractor had such a broad smile on his face when they were awarded the bid. Bonnie still believes she actually heard some snickers and giggles from the other bidders.

So now they had their dream project and they were off to the races. Everything looked good on paper, and heaven knows they needed the work. They had no jobs lined up at the time and needed money. But what seemed to be their ticket to the big time soon turned into what they "affectionately" call their ticket to hell.

Although they had attended the Estimating Workshop, they had not fully grasped the idea of overhead recovery and mark up. As a result, when they bid the job, they added ten percent to their cost and that was their bid. Nothing additional for possible over-time, bad site conditions, overhead, etc.

The job had started out pretty decent and ended in a nightmare. What was supposed to take three months actually took six months. Plants were supposed to be delivered at a certain point ended up sitting on top of the asphalt because suppliers and de-liverers could hold off no longer and as a result a larger percent-age were destroyed and had to be repurchased. The dream was crumbling at an incredible rate!

Things had regressed so terribly they were politely advised to de-clare bankruptcy - quickly! But having a "vast storage of pride," Ron and Bonnie felt they could not take that particular route. They were warned by the posh legal offices not to fight it, that four out of five businesses would fail in their position. They were stuck between a rock and a hard place. One thing was for certain - they did not want to let down their suppliers, but were afraid to admit to the world how much trouble they were really in. They finally agreed to the old adage where honesty is the best policy. They would tell their suppliers the truth and attempt to work things out - without the bankruptcy.

After several attempts, many phone calls and a large degree of frustration, they were able to receive an SBA loan. It was not enough to pay everyone everything, but it was an extremely large step which proved to be vital in their survival. Bonnie immedi-ately contacted all of their suppliers to make arrangements for payment. All but six of their suppliers agreed to wait for their money. After more struggling and juggling, they were able to complete the job, pay off the rest of the suppliers, but most im-portant of all, retain their pride.

The previous winter had not been extremely kind to Ron and Bonnie from both a business aspect as well as financial aspect. They then decided in order to get back on track they would enter into the general contractor low-bid market. They were well liked by the generals due to their reasonably low bids and they produced some exceptional projects. It provided them with cash flow as well as got them noticed in the local community. Both Ron and Bonnie will be quick to tell you they believe this is the only thing which really helped them to survive. Their good work began to pay off when at later dates, the generals who had taken notice of their work began to call them back.

By the fifth year in business they had managed to salvage the company, had moved into their new home and life was looking very cheery - so they thought. They had been told if they could survive through the fifth year, they would have it made. Buy by Christmas of the fifth year, they had found that to be false. The past 4 ½ years of mistakes had finally caught up to them and they suddenly realized they were $300,000 in the hole and were now truly facing bankruptcy, and possibly even going to lose their house. They had employed a financial advisor at the time they had received their SBA loan, but it was all too little too late. The IRS began pounding on the door demanding back payroll taxes in excess of $50,000 or they were going to change the locks on the doors. The office manager had quit because she couldn't take the stress of the nasty phone calls. No Christmas cards were sent that year and no bonuses were given. Their second baby had just started to sleep through the night - but they couldn't.

They had thought of just giving up. It would be impossible to pull out of such a mess and have anything left to survive on. But they didn't give up. Shortly after this disastrous Christmas, two large projects came through for them which were totally negotiated and for the first time, relatively profitable. One of the developers paid them deposits so they could locate their materials early. This in turn saved him money, which went into their pockets and enabled them to pay the IRS. They quit working with general contractors and began working with owner-builders only.

Because of the excellent work they produced, they began getting several calls for specialized work which other companies were not doing. They also entered the high-end residential field and slowly the red in began to fade back to pink. By this time they had also attended another Estimating workshop, had drastically reduced their overhead to a reasonable number and were charging customers more than just the COST to do a job.

It is eight years later and Ron and Bonnie are still in business. They are still recovering from their past mistakes. It has been an extremely rocky road, both personally and professionally, yet they persevered. Bonnie does not believe any of their success came in the form of luck. She will gladly admit maybe a "Higher Authority" looked down on them at one time and said, "Let's give them a break. They are good kids and they mean well." But she believes the main reason they survived was because of pride and determination.

She stated the one important item for them was they never gave up. Although they didn't have any money, they ALWAYS made sure they finished the job they had started and always produced the best product possible. They never cut corners or skimped on items, and always insured the customer was satisfied before they left. This paid off in triple dividends when customers began calling them back for more as well as referring them to other people. Proof positive good ethics, moral standards and integrity pay off!

Bonnie will also be quick to tell you how important it is to know the "business end" of landscape contracting. She no longer hesitates to collect money owed to the company. She has learned the proper way to bill receivables (and collect them) as well as charge (estimate) the proper amount for every job. She says one of her biggest problems was that she would actually begin to feel guilty about charging clients too much money - money the company needed.

Bonnie will also attribute their survival to the fact they faced the problem. She will readily advise ANYONE who becomes involved in a financial crisis to face the reality. She is very quick to tell friends, clients and even other contractos that playing OSTRICH does not solve the problem. Because when you pull your head out of the sand the problem will still be there - along with the IRS, collection agencies and your suppliers.

FOOD FOR THOUGHT

Some Ideas Which May Help

As we traveled the United States and Canada, we interacted with hundreds of contractors. Many times they shared some common problems and frustrations they had experienced in operating their various businesses. Sometimes they shared ideas and methods they had discovered to solve them. This section provides us with a format to discuss some of these problems and frustrations and explore some possible solutions.

Not all of these ideas can be used by your company; however, if you can get one or two or even three good solutions and implement them into your company's operations, it could save you many times the price of this book. As always, we recognize the wide diversity of business people who are reading this book. Consequently, we do not guarantee the success of any idea for any business. We encourage you to use your best judgment as to how well any idea would work in your particular situation.

By purchasing this book, you have shown your willingness and desire to change and make your company better. This attitude, by itself, will give you a better chance of survival than most others.

Bad Jobs

For a year, in over 60 speeches before contractors in the United States, Canada and Australia, I introduced a new campaign. This campaign was to increase profits by 300% through the implementation by each contractor of the ideas presented in the campaign.

A contractor can survive if about 20% (about one out of five) of his jobs go bad. Now by "bad" I mean break even. He bids a job for $22,324 and it costs him $22,324 for the materials, labor, equipment and subcontractors to do the job. It has not cost him any money out of pocket to do the job, but there isn't any money for overhead or profit. If he has any more than 20% of his volume or jobs to this way, he cannot survive.

Here is why. It takes the profit of 60% (three out of five) of his good jobs to pay for the overhead on one job. If a contractor were to do $1,000,000 in work and $200,000 went bad, it would cause the loss of profit on $600,000 worth of work just to pay the overhead on the one bad job and the only profit realized would be the profit on the remaining $200,000 worth of work.

So often I find contractors who do not look at a bad job in the way I just explained. They think if it did not cost them anything out of pocket they can go on to do more profitable jobs and still show a profit. As you can see, it just does not work this way. So here is the campaign everyone needs to take on: CUT YOUR BAD JOBS IN HALF!!

Bad jobs generally happen in two ways. First, they are improperly bid. From the very beginning they are bad because your bid has mistakes in it and the contract price will only cover your costs. Often the quickest indicator of this is the difference between your bid and the other bids. If you are more than 10% to 15% lower than the other bidders, your bid is probably wrong. When I found myself in the situation where my bid was more than 10-15% below I have even tried to convince myself in those situations I was right and everybody else was wrong. However, in 90% of those cases it was just the other way around and I was hung with a break-even or loss. This isn't guaranteed to be a good indicator because I have experienced my bid was less than 10% low, yet it was still a bad bid (even though it looked good in comparison to the other bids). This was because other bidders had also made mistakes. My mistake was just bigger and made me low bidder.

The best way I know to keep this from happening is to spend lavish amounts of time on your estimating and bidding process. If you are not on a good estimating system (not necessarily a computerized system) you absolutely need to be because it can protect you from many of the unusual circumstances which contribute to a bad bid. You also need to know how to review a bid and look for some of the quick indicators which show you if a bid is questionable. Those indicators are given in a subsequent chapter called "Bid Reviews."

Another general way jobs go bad involves all the events which occur once you sign a contract and get a job started. Other trades may be in your way and keep you from performing efficiently. Unknown site conditions prove to be difficult and cause you problems. Owners or their representatives become unreasonable in their demands and hold you to every letter of the specifications. On and on goes the list but you have the idea. The only way to avoid those items is to gain that sixth sense which detects those things before you bid a job or sign a contract. Again, spend lavish amounts of time investigating the site, the owner or his representative, and the reasonableness of their proposed schedule. Talk to others who have worked with the owner or the architect. Develop a system of critiquing your thoughts about the job and all of the information you gather about the job.

If you succeed in cutting your bad jobs in half, let me show you how it will increase your profit 300%. Instead of having 20% (one out of five) of your jobs go bad, only 10% (one out of ten) of your jobs go bad. It will still take the profit of three good jobs to pay the overhead on the bad job, but now you will have six out of the ten jobs' profit left instead of one out of five or two out of ten. Six out of ten jobs' profit as your total is 300% better than two out of ten.

Bad jobs will always exist in this business. There is no way to completely be free of them. However, let it be your campaign. CUT THEM IN HALF.

Bid Reviews

In the previous idea, we talked about bad jobs and how to reduce them. In this idea I want to give you some thoughts on how to review your bids in order to reduce the number of bad jobs you get because of bad bids.

I want to set the stage for thinking about bid reviews by saying that, for some of you, a second party (other than the one who did the estimate) can review the estimate. For others the only one to finally review the bid is the same person (usually you) who did the bid. If this is the case, I still recommend the same procedure, the only difference being when you are done with the bid, walk away from it for a while and when you return, put on a critic's non-biased attitude toward the bid.

The first thing I do when reviewing a bid is to skim through the specifications and plans. I develop questions to ask about the job, some general in nature, some very specific and detailed in nature. I discuss these questions with the bid writers. I not only want to receive the answer but to asses how confident this person is in the answer. If they are hesitant or they seem unsure about some items I know I must dig deeper into the plans and specifications, because the bidder may have missed some things. I also want to know if the costs on these items in question are in the bid.

Next, I look at the material-to-labor-to-equipment ratio. This is, in most construction disciplines, a fairly standard ratio which exists for these items for certain kinds of work in your company. You should know what those ratios generally tend to be. If a job is outside those normal ratios, it is time to ask some questions and look into why this is so. It does not necessarily mean there is a mistake, but there could be and it deserves further investigation.

This investigation leads me to the next place I would look, which is the production rates for labor on the job. When I look at production rates, I generally skip past the standard items which are done every day and look for any new types of work or items which might be on this bid. I then ask myself or the one who did the bid to reason with me as to how they came up with those production rates. If there is disagreement on the production rates it may be time to call in a field person to get his views on how long certain types of work may take.

Then I focus on the materials for the job. Sometimes the ratio is out of whack because there are unusually expensive or inexpensive materials on the job. I also want to know where the BIG money items are coming from and whether the supplier can be trusted to hold firm on his price. It is also important to know if the prices include shipping and whether or not there is a potential problem in getting delivery in a reasonable time.

Next, I look at the subcontractors. I look at three critical areas. First, are they subcontractors which we have used before, or, if not, do they come with good recommendations? One of the best ways to answer this question is to know if they are substantial enough to supply a performance and payment bond for their portion of the work. Second, how does the low subcontract bid compare with the other bids which have been submitted. If it is more than 10% lower, I want to carefully look at the third factor which is their inclusions and exclusions. In order to compare their bid with others (when they have excluded more items) I may have to figure what those exclusions would cost and add it to their low bid to see if they are still low.

An important questions has long existed which is, "If a subcontractor is ridiculously low, should I call him and tell him?" I have always answered with a resounding "YES!" I would never tell him how low, or that I expect him to raise his price. I tell him I am calling as a courtesy and showing my concern because there is a difference in the prices and I want him to look over his bid one more time to make sure everything is right. I hate to use a ridiculously low sub bid when I know if it remains out there someone else may be crazy enough to use it and ruin the potential of me getting the bid.

During all of these checks I am usually doing one additional thing. In my mind I roughly calculate some of the largest numbers on the bid to make sure there is not a mathematical mistake.

Then I look at the General Conditions. As a rule of thumb, they should not exceed 10% of the cost. Again, there are no absolutes in this business, and I have seen them exceed this. But if they get above 8% of the cost of the job, I want to look at them to make sure they are not excessive. I also want to look at them to make sure all the typical items and the non-typical items are included. Next, I look at the overhead recovery and discuss the profit. These items are discussed at length (such as how they're calculated) elsewhere in the book (see page 465). At this point I want to total the overhead and profit on the job and ask myself this question, "Is this enough money considering the size and/or risk, and the effort which will be required to do this job?" There have been times when I used the right overhead recovery numbers and reasoned a good profit, but after considering this question, raised or reduced my price.

Finally, if it is a unit price bid I scan down the final unit prices. I have, or should have, a feel for what things are priced at in the local market. I look for items which are higher or lower than normal at this time. Again, this does not mean I change the price, but I must look at it again to see why it is out of line with the normal price.

None of these things will guarantee a perfect bid, but by doing them you will reduce the number of bad jobs you get due to bad bids. I have found the more you do this, the more you will acquire that sixth sense which will warn you of a potential problem.

Billing "Trigger"

Although probably the shortest of all ideas contained in this section, I feel this is probably one of the most important, if not the most important idea for a company to implement. One of the concerns which can often haunt a construction company is whether they have billed all the work they have done in a given month. A contractor we know has set up a "trigger" system which will begin the billing process on a monthly basis. Each week (or month) the company owner has the accountant generate a list of every job for each particular month which has had labor performed on it. This list is taken from either the payroll reports or from the job costing sheets module.

This list is then compared to the Accounts Receivable list to determine which jobs should be billed and for what amounts, and an invoice is prepared immediately and sent to the customer.

There are two very important steps which need to be taken in order for this system to be efficient. First, you should have a good job costing system in place, whether it is done by computer or manually. Second, your secretary/accountant must be extremely efficient in order to prepare this information in a timely manner. If you do not have in-house accounting, whether manually or by computer, consult with your Accounting/Payroll service to determine if they will provide you with the necessary information. Not only will this system assist you in your budgeting/projections, it will also keep money from "falling through the cracks" due to jobs which were not billed in a timely manner.

Bulletin Boards

From our earliest days in school we have all been around bulletin boards. Usually a company has one which is supposed to be where important information and government bureaucratic notices are hung for everyone to read. In most cases, they become quickly ignored or looked upon with complacency.

A contractor we know has revitalized the bulletin board concept as a good method to communicate with his people as well as build their self-esteem. He has given each of his key personnel or foremen a bulletin board displayed together in a high traffic area of his company. The person's name is boldly put on the board so everyone knows whose board is whose. However, to make the board mean something, he says three things must be considered and worked on in a continuing manner.

First, make the boards interesting. He accomplishes this by putting up some of the following types of things: Any letter of commendation by clients or anything which can be considered an "At-A-Boy"; jokes or comic strips which might apply to a particular person or something they have come across; memos from you which compliment or show appreciation for a job well done or a good attitude shown or a time when they went beyond the call of duty.

Second, use it as the only place where they will find important communications from you to them. It can also be used by each foreman to notify other foremen they are in need of a certain piece of equipment or advice on how to solve a particular problem. They would simply need to create their own memo or request and post it on everyone else's bulletin board. But notice the words "only place" in my first sentence. If you use some other box or place where you will leave them notes or written instructions, then they will soon pass up the bulletin board for this other location and not notice the things which are on the boards.

Third, appoint someone within the office to keep the boards semi-clean. If things stay on the board beyond their useful interest life, they will begin to cause people to ignore the more current and important items which have been placed there. You might even use a system in which the person initials some piece of communication or information, thus acknowledging he has seen it so it can be taken down.

Some would feel using a bulletin board can make too many items "public information." However, the contractor who uses them feels this is the exact purpose he wants served. He feels certain things should known by everyone and he actually encourages people to read each and every bulletin board. This way, if someone is in need of some help, others will see it and be able to help. He also feels it helps to create a feeling of openness in the company without using the unreliable and often false "grapevine" as a method of getting information to all employees.

Maybe it is time for you to reinvent the successful use of bulletin boards in your company. However, a word of caution. In the beginning of this section we talked about placing jokes and comic strips on the boards pertaining to certain people, jobs, etc. But, please, always keep in mind you are dealing with different personalities within your company, and what one person might find humorous, another may take as an insult. Remind your people to always use discretion before placing any items on the boards.

Burnout

Burnout is certainly not a new idea, nor is it an idea which anyone would like to implement in themselves or in their company. However, it is a fact of most contractor's lives. In fact, one of the most prevalent problems I face with the owners of construction companies is this very reality. I have come to believe somewhere between the twelfth and sixteenth year, someone who is on the front lines of this business will face burnout. I want to talk about it here because some of you are feeling it and may not have any idea how to get out from under what you are feeling.

Burnout has several symptoms for a contractor. It often is manifested in a loss of excitement for the projects you are doing. You begin to lose the excitement and challenge which contracting offers. You may find it is hard to get up in the morning because you are not as excited as you once were toward the company and the projects you are contracting. New bids and even the successful awards of new jobs bring a pessimistic feeling and even a dread. Another sign is you may find yourself losing patience with the people who work for your company. Their problems and the mistakes they have always made are now becoming monstrous and you are getting sick and tired of them. Another sign is when you begin to mistrust every client, architect and engineer for whom you are constructing projects. You constantly believe their every move is motivated by a desire to get to you and your bottom line.

These are just a few of the major signs you may be getting burnt out. The problem is not these problems or the problems in this business, but rather being able to recognize burnout and do something about it before it is out of control.

Several of my clients have given me ideas as to how to overcome the burnout and to come out it feeling revitalized and rejuvenated. One client likes to get out of the office and actually go supervise one of his jobs. I have had clients who have actually gone out and done this for a four-month period of time, having turned a great deal of responsibility for the everyday functions of the company over to other company members/employees. Others have taken time to involve themselves in some form of outside volunteer work in order to be able to collect their thoughts about their own company.

Several of my clients have found it wise to gain a mentor. This person usually lives and owns a company in another state. They may go and visit this person for several days to confide in them about their struggles and work through their feelings. We have held several meetings throughout the country every year which have become something of a "support" group. Contractors come to these meetings to sort through their feelings and hear from others how they are not alone in their struggles.

Whatever the symptoms, whatever the cause, whatever the solution, the message I bring is the same. Each of you, in one degree or another, will face the problem of burn out. When you do, you are among the vast majority of people in this business. Burnout is not the problem. Recognizing it, admitting it and then doing something about it will make the difference in your ability to carry on successfully in this business over the long haul.

Collections

A Canadian contractor shared a unique idea he has used to deal with collection of money from some difficult clients. It may not work every time, but it may be worth a try in certain situations.

When someone is dragging his feet and not paying, and is using excuses which do not make sense, the contractor finds out who his attorney is. He then goes to that attorney, acting like he is a potential client. Without using the name of the person or company not paying its bill, he describes the situation. He may even show the attorney a copy of the contract with the client's name blanked out.

After describing the situation and answering any questions the attorney may have, he asks the attorney if he thinks he has a legitimate case for collection. If the attorney feels he does, he is asked to start a suit against the client and then (and only then) reveals that client's name.

When the attorney realizes it is one of his clients, and informs the contractor it would be a conflict of interest for him to take the case, the contractor plays dumb and says "but you feel I have a winnable case because he is in the wrong." After the attorney still declines to take the case the contractor leaves, but many times discovers the attorney has called his client to tell him two things. First, you are looking for an attorney to help collect the debt, and second, he feels the debt is legally collectible and the client does not have a chance of avoiding collection. The contractor reports, just after the trip to the attorney, the client has paid the bill.

I have also discovered one of the most important words to describe your attitude about collections is the word "urgency."

Everyone who is called or approached should be approached in a manner which indicates the money they owe you must be collected immediately in order for you to survive. Never be afraid or ashamed to take on this attitude of urgency when it comes to collecting money.

Contests

When you start a business, from the outset you must determine the company culture and approach to your employees. Are you going to run a company where your employees work for the company as an employee? Are you going to pay them a set salary and try to motivate them to do a good job for the sake of the company? Are you going to motivate them to make a profit for the common good of everyone in the company? Or, are you going to run a company where you encourage your employees to not be merely employees, but small companies within the company? You pay them commissions or distinguishable incentive bonuses or subcontract work to them. You build on their private entrepreneurship desires and skills and set them up in business within your business.

Well, no matter which one you choose, a good way to make things happen in your company may be through use of contests. I like contests for the following major reason: they fulfill the desire of everyone to gain something from their talents, ideas and abilities WITHOUT becoming a long-term company policy. Everything you do in a company must be evaluated based on the kind of precedent it sets, or on the long term effect it will have on the direction of a company. Not so with contests. They stand alone based on the rules and purposes of each individual contest. Whether you run one or two a year or one every five years, that is your prerogative.

Let me tell you of an all-encompassing contest which I heard one contractor ran at different times with great success. I call it all-encompassing because it can bring great results to any or all parts of your business without being too specific. He called it "The Idea Contest" and here is how it worked.

Every employee (except the judging committee) was encouraged to enter the contest. Within a six week period they were to submit, in writing, any and all ideas about how to improve the company and make more money. As you can imagine, ideas about everything from equipment maintenance to paperwork to management procedures to production ideas came forward. Once they were submitted they became the property of the company to use or not use as they saw fit.

A committee of three judges was established to look over the ideas. They were to choose the best three ideas from a practical, functional and profitable perspective. He awarded the submitter of the top idea a seven-day cruise for two people. The second best idea received a one-week vacation for two at a destination resort. The third best idea received a weekend for two at a resort area within driving distance.

Even though he spent over $5000 for prizes, this contractor has always felt he has received ideas which have made and continue to make him a lot more money than he has ever spent on prizes.

Customer Profiles

I read a very interesting book which had an idea I feel is worth passing along to you. The title of the book is "Swim With The Sharks Without Being Eaten Alive" by Harvey Mackay. It was sent to me as a gift by a contractor friend of mine and I highly recommend you get this book.

In the book, Mr. Mackay has a customer profile made up of 66 questions which he ask his clients and "would-be" clients to fill out. The answers give him a profile of his customers so he can talk with them about their families, hobbies and even what their favorite foods are so he can take them out to eat.

Many of the questions on the list are not appropriate for the type of clients you have in this business. However, there are several which are. If you get a copy of the book, you can pick and chose which ones you feel are appropriate. But if you don't, I want to give you a general list of the kind of things you need to find out about some of your steady clients. This will give you a good insight into their likes and dislikes, and give you the ability to converse with them on a broader spectrum.

There are three areas in which I would suggest you build a profile for your future reference. First, family. Find out things such as family members and ages; family member involvement in sports, hobbies, school or other important ventures; birthdays and anniversaries may also be good information to have.

Second, personal. Find out a person's likes and dislikes concerning food, sports and/or recreation activities. Also, such information as education, military experience and religious backgrounds can be of benefit to know before dealing with a person on a regular basis.

Third, business. Find out how long they have been in business, what their goals are and what are some of their greatest satisfactions and frustration in business. Such questions can give you the ability to know how you can best help them with your business.

Some people just prepare a questionnaire and have the client fill it out while others fill out over a period of time, by asking the pertinent questions and filling out the information themselves. Whichever way you do it, we should strongly suggest you have a file which contains this kind of information for certain key people you deal with on a regular basis. This is a people business, both from the aspect of the people working for you, as well as the people you work for. People like to know you know them personally and have taken the time to know something about them. It can be a great advantage to have your sales people review the file you have created on these potential clients, before they go for the sale. Also, instruct you sales people to write down or bring back any additional information. Your best bet is to ask the questions in a very informal and relaxed type atmosphere. Keep in mind you are there to try to get to know the individual, not conduct a job interview.

Delegation

There is in this business a plateau that many companies reach which is call "Founder Gap." It is the place where the company has grown in size and profitability to the "Founder's" limit of performance. In order to get past "Founder Gap" to additional growth in sales and profit, an important thing must be done. This important thing is delegation. Consequently I want to share with you some things about delegation.

Usually one of the first things which must be overcome is the founder's EGO. He has built his business and has done it all. To this point, the success of the company is a direct result of his efforts. Often there is a strong feeling that if it is going to be done right, the "Founder" must do it.

I have heard of three important steps which must be applied if delegation is going to work in your company. First is accountability. When you delegate something to someone, they must be made accountable to accomplish that which you have delegated. They need to be explicitly aware the "buck" will stop with them and there will be no excuses or finger pointing accepted.

After you have established accountability and the person has accepted it, then comes the next step, education. Too often we assume just because we delegate and because people accept responsibility, they must know what they are doing. I have found, more often than I want to acknowledge, this is just not true. Many times people accept things and they are way over their heads. They have accepted it to please you, or to grab at the opportunity for advancement. Consequently, with delegation must come your willingness to educate this person and make sure they have a running chance to succeed.

Finally, there needs to be verification. I believe it was Tom Peters who said "What gets measured, gets done." Everything of any consequence you delegate must have a method by which you can measure its successful completion. Too many contractors are naive enough to think their employees are as conscientious as they are. If they were, most of them would be contractors themselves. Consequently, I have learned a lot of what I delegate will not get done, and I won't even know it unless I set up a system of verification.

Once I delegate something to someone, I put it on a list of things which have been given to them. As often as I can, I go over that list with each person to see and verify it has been either done or is going to be done.

If you are frustrated with delegating things to your employees, I doubt it is their problem. Most likely it is YOURS. You are not following these basic steps: Accountability - Education - Verification.

Design and Build Contractor Plans

Design and build contractors have many advantages over those who must strictly compete on the open bid market. In almost every case I have worked with, those who are able to do some design and build work find work is almost always more profitable, better looking when done, and less of a problem to actually build. However, after talking with a design and build firm, I feel I need to pass on a few tips which many of you may not be aware of in what you are doing.

First, because you are a contractor, the planning of any new project should take into consideration the problems which will or could occur during construction. All designs should be looked at not only from an aesthetic perspective, but also from an ease-of-construction perspective. This very concept is one which you need to use in selling the worth of using a design and build contractor in the first place. Many field problems which come up from a design done by a non-contractor should not occur in a design and build environment.

Second, all plans which are produced should be plans produced for construction, not for show. Renderings and sketches can be too "showy" and dramatic for the purpose of getting across ideas and concepts. However, when it comes time to make the drawings, your clients are more impressed with drawings which show the details and have the explicit notes which are necessary for your field to actually do the work. Remember, the strength of your relationship with the client is you are more than a designer, you are a contractor who knows what is needed and what will go on in the field.

It is these two points where the profitability of a design and build firm depends upon. If you are not successfully doing these items,

design and build work will not be as rewarding as it can be for
your company.

Finishing Jobs

I have developed new terms for the personality of contractors. In another chapter of this book titled "Four Personalities That Must Exist In a Company," I describe four distinctive personalities which must exist and work together in the management of a construction company. One of them was called "A Builder," someone who takes another person's idea and builds it. I now call that person "The Happening Man," which is someone who takes another person's ideas and make them happen.

But with the personality of "The Happening Man" often comes an inherent problem. Many times they are such aggressive, get-to-the-next-thing people that they are not good finishers. They are not good at seeing and performing all of the little final details which need to be done to finish a job. Several of our clients have identified this problem with many, if not all, of their key, high-production foremen or supervisors. They can get a job going and smoke blows through it until the last 10% of the job. When they get to that point they flounder and begin to lose money and often they never get it all done right.

We suggest you consider this idea if you have some of these field people. Rather than try to push a rope uphill and make them something they are not (i.e. "A Finisher"), develop a finishing crew. For every two, three or four crews you have starting and doing jobs, have one small crew which specialized in coming in and finishing a job.

When we were building golf courses, for instance, we had "A Finisher." We had people who could go in and move great quantities of dirt, build greens and put in massive irrigation systems. They could get it within 90% of the way it should be. Then, we would pull them off the job and send them to start another one and bring it to 90%. When they were gone, we brought in the "Finishers" - the people who could see all of the little things which needed to be done for completion and get the job finished and looking good. They had a sixth sense and an eagle eye to methodically and consistently go through the remainder of the job and finish it.

Hiring People

In most situations, hiring people simply means putting an advertisement in the local paper and screening all the many applicants who come to apply for the job. However, in certain economic times and in certain parts of the country where the economy is strong and business is booming, few people may turn up for such jobs. I have heard, in certain parts of the country, people were paid big wages to flip hamburgers and so construction labor was considered too strenuous for the money offered.

We heard of the following idea for getting labor in those situations and keeping them with you. A contractor advertised in the local papers under construction labor and added this concept. Anyone who came to work with him and worked at least eighteen weeks received an all-expense-paid trip to a destination resort.

The contractor kept the expense of the trip to around $600 to $750 and found because he was buying in quantity he was able to have his section of some nice spots. He did not go into great detail about the rules for the trip or of what would happen if a person was fired prior to the trip.

However, he did have some guidelines. Workers could be fired or quit within the first six weeks without having any portion of the trip awarded. After six weeks and for the next twelve weeks, they were accumulating one/twelfth of the cost of the trip every week. If they were fired or quit during that time, they were given a cash bonus equal to the amount accumulated. Just prior to the contractor having to book the trip with the travel agent, workers were also given the opportunity to not go on the trip and take it as a cash bonus, or pay for additional people to go with them on their trip.

The contractor stated there were two nice extras in this idea: first, he and his wife got to go on a company paid trip; second, it helped build camaraderie among his employees.

If getting good people to respond to your advertisements is becoming difficult, you might give this idea a try.

Human Resource Person

I have known of the existence of these kinds of people in the construction business, but I previously thought they were found solely in larger companies. However, one person I know in the industry, with a company in the three-million-dollar range, has had one as part of his staff for over a year and thinks it's just the ticket. So, I want to explain what this type of person can do for you and to recommend you think about hiring such a person (or at least to give one of the people in your organization some of these items as part of their job description).

A human resource person can or should do all or part of the following.

1. Hire and legally terminate all personnel. The contractor found this saved him lots of time and energy as the human resource person weeded out a lot of unqualified people. They found the people who had been doing the hiring before could now better concentrate on their main focus. They also found they were getting better-qualified people.

2. Listen to employees and act as a screen between all of their little complaints and wants, and the owner who usually would have to handle those things.

3. Help establish company policy for employees. Because of his knowledge, what people were telling him they would not tell the owner.

4. Check into the benefit program of the company. This meant the pricing of the best insurance program for both the people and the company, as well as knowing

when people qualified for vacations, insurance and any other benefits.

5. To be responsible for a good all-around training program for all of the employees. He was to set up such training with the help of field supervisors, and screen the myriad of seminars and college-level programs available for management and recommend which they should attend.

6. To monitor the company's EEO and safety programs to make sure they were in compliance with government requirements which applied to their company.

This is not intended to be a complete list of what such a person could do, but it should give you an idea of what they could perform for your company. No matter what size your company is, you might consider such a person or the assignment of some of these items to just certain people, rather than a lot of different people.

Indicator Of Economy and Your Success

What exciting things the computer age has brought to this business! I am constantly amazed at all the things which can now be done quickly and economically which just ten years ago were out of reach of most contractors. One of these things which a contractor we know has been doing has proven to be very beneficial to their forecasting and budgeting.

In a spreadsheet program like Lotus 1-2-3, he has been tracking the number of leads which come in each month by telephone. He then keeps track of how many of these leads they are actually able to follow up on and turn into proposals. Finally, he keeps track of how many proposals (by salesperson) are actually contracted and completed within the given year.

All of this took some extra work on the part of his secretaries as well as himself, but when we sit down to look at his year and the year ahead, these reports help us better understand what is happening in his company. For instance, we found one salesperson who was taking a draw was not nearly as successful as another who was taking the same draw, so the owner could take appropriate action. We also knew how many more leads needed to be generated to reach his new sales goal for the next year. Also, during the year, he has the ability to know when the salespeople are hitting below their average percentage of leads to proposals and proposals to contracts so discussions and changes can be made before it too late.

Keeping Track Of Crews

One of the things which is always a problem with field people is paperwork. The greatest of the hated "paperwork" is any job costing or time card reports which have to be kept. This problem becomes even greater for maintenance or tree trimming, in any type of service company where your crews are on many jobs during a one day period.

Here is an idea which can help keep track of these crews as they move from job to job. It can even be used by a foreman on a big job; however, this idea does require you have a radio communications system in your vehicles. One contractor has the foreman radio the office each time they arrive at a job with a simple code, i.e. "Foreman #10 (example) and number 1 for arriving at a job name. For example, the foreman would simply say, "10-1 at Smith residence." At that point, the secretary monitoring the radio acknowledges the message and logs the time. When the foreman is done and ready to leave, he radios "10-2 (2 for departing) the Smith residence"

At the end of the day, you have a log of the hours spent on each job and travel time between each job. The office personnel can either do the "hated" and often incorrect paperwork for job costing. This system can also help minimize those "extra" stops at donut shops, lengthy lunches, etc.

Meetings - Listening

One of the things I have been emphasizing to contractors around the country for many years is that this is a people business. Our clients don't call us to order material (although they could do that). They don't call us to find subcontractors (and they could do that, too). They call us because they want us to supply people to install the materials and "construct" a job.

I have further stated if this is a people business, one of the most important skills we can develop is not just construction skill but also people skills. I believe one of the most important skills which needs to be developed is the skill of communication. When I talk of communication, I emphasize one of the most important parts of our bodies we use to communicate is our ears. Listeners are good communicators and one of the most important reputations you can acquire for yourself is the reputation of being a listener. You cannot believe the thousand of dollars which can be made by simply listening to people.

I tell contractors this as a consultant. I have been paid my fees to simply come into a company to find and solve their problems. Many times I go to their employees who tell me of the problems and they tell me the solutions to these problems. I then sit down with the owner and tell him his problems and the solutions I got from his employees. It is something the owner could have done without me, if not for one major problem. The employees would not have told him what they told me. Do you know why? Because he was not a listener and the employees know it so they quit talking to him other than in essential conversation. All of the ideas and solutions to problems were there, but not being shared because no one would listen.

With this thought in mind, I want to share an idea that a contractor I know is doing in his company. Throughout the year he conducts a series of meeting he calls "listening meetings." Their entire purpose is to get with certain people and let them talk and share their frustration (problems), their ideas and, eventually, the solutions to many of their frustrations and problems. He does this with all of his employees several times a year. He also gathers some of the architects to do the same thing. Many times he does it over a lunch he puts on for them in a local hotel meeting room. He also has done it with his clients, whether they are developers, general contractors or subcontractors. He also has done it with material suppliers and/or anyone else when he feels can bring him information as to how he can run his company better and make more money.

Take time, in some special format and setting, to listen to other people and you will find yourself more educated and more successful.

New Business

There has always been a problem for contractors who own a lot of their own equipment. The problem has been the maintenance of the equipment in a cost-effective and hassle-free manner. Some have hired their own mechanics and tried to justify them and/or used those mechanics out in the field if there was not enough work for them to do on equipment. This is a method of operation which will surely cost you a mechanic or two.

Others have tried to get "married" to the repair shop in hopes of getting better prices and better service from such an arrangement. However, the repair shop views this type of situation as very dangerous for them because the contractors can build up their shop to service the contractor, only to wake up some morning to find themselves "divorced" and him going down the block to "marry" another shop. Consequently, the shop can never quite give you the service or prices contractor may desire.

One contractor who ran a business in which he had a lot of small equipment with small engines, came up with the following idea. He would open a shop as a separate business, with a different name, right next door to his shop and office. He would advertise and take on the repairs of small engines (bobcats, lawn mowers, chain saws, compressors, etc.) as a business. The business his construction company would give this new business would keep it busy more than 50% of the available shop time and the general public would bring in the rest, giving him the luxury of having his own shop next door while making a little money from the outside business.

Newsletters To Clients

Elsewhere in this book, I talk about having a newsletter to your employees under the subject of the "scoopline." In this thought, I want to talk about newsletters to your existing and/or potential clients, as well as architects. I have seen several clients use a newsletter effectively. They go out from time to time, and really promote the company in an interesting way. Every time a newsletter goes to someone, they are once again reminded that the company is out there in the marketplace.

However, I have seen many companies start sending newsletters to clients but they are unable to maintain a consistent program or finally quit sending them. They usually quit because it becomes too big a job. This becomes more debilitating than if they had not started, because those who used to receive the newsletter now wonder if something has gone wrong with the company. Or, they view the company as probably incompetent because they start things but are unable to finish them.

Other companies want to start one, but don't know how to get it started or realize how big a job it is and are too overwhelmed to get the job done.

One idea is to use a prewritten newsletter base which may be purchased from a newsletter source which is about 80% written with standard, interesting information. About 20% of the newsletter is written by someone in your company with information specific to the company. Anyone who wants to have a newsletter contracts with the special source which produced the bulk of the newsletter on a regular basis, monthly, bimonthly or quarterly, You produce the balance of about 20%, adding the company logo, name and address. This way, a newsletter can be produced by your company and sent to your clients, prospects and architects, with only 20% of the work falling on you.

To search for such a newsletter provider, check ads in trade magazines, making it specific to your industry, or check the Thomas Directory at your library for newsletter sources nationwide.

Pre-Qualifying Customers

I tell contractors one of the things which scares me the most is millions of dollars goes through their hands. What scares me about this is often contractors think more of the money is theirs than really is. So, do you know what they do with it? They spend it! Then the real owners of the money come for it, usually the IRS or material suppliers.

The same concept holds true with the customers for whom you are doing construction projects. They go to the bank and borrow thousands, if not millions, of dollars to do a project. As they are doing it they begin to think they have more money to spend on this project than they really have. As the project goes along, they find the available funds becoming less and less. So guess who doesn't get paid like they should? The contractors.

I have heard of a general contractor who would shut down his major projects the first time it snowed. He would then go to Florida and hole up for the winter. Before he left, he would deposit his last payment in high interest bearing accounts, not pay his subcontractors their last draw, and leave a secretary in his office to explain he was out of town and would take care of it when he got back. He didn't get back until March and he lived off the interest of those final payments he did not disperse.

How do you keep from becoming a victim to these and many other types of nonpayment in which people live off of your money, or use you to finance their development? Here are some ideas which contractors have shared with us.

First, talk to some other subcontractors or contractors who have worked for this person or company. Contracting tends to be a very secretive business. Contractors feel they cannot talk to each other or they will give away some great trade secret. This attitude, I feel, comes from the concept of "competitive bidding," a term I have been trying to dispel in this business. The fact is because this is a very difficult business with some of the world's shiftiest crooks in it, the good contractors need to be in open communication to keep from having some of these crooks prey on them. I know a contractor who goes onto a project before he bids it and talks with as many contractors as he can with these questions. "How does he pay?" "How easily are change orders considered and handled?" "How reasonable is the owner's representative or project manager in his requests and interpretation of the contract documents?" I have also found my bonding agents very helpful with answers to these questions, because usually they would have heard if another contractor had problems with the owner on a previous project. Also, in certain cases, material suppliers may be a good source of some of the same information. They have gained it from first-hand experience, or from other contractors who are not paying them on time because of problems they have had with this developer or owner or architect.

One thing is certain, I have always discovered an important part of my job was to be a snoop. I found the more questions I asked, of as many people who would answer them, the more I kept myself out of trouble.

In every larger city there is yet another source of this kind of information. It is the Subcontractors Credit Association. Check around in your area for this group and consider becoming a member so you can keep yourself better informed before you start a job.

There are two other ways which will help pre-qualify an owner or developer before you go to work for him. Both require him to produce information and/or proof of his ability and willingness to pay for the project once it is completed. If the owner or developer is not willing do these two things, you should begin to question his credibility.

First, you can ask for and obtain a letter of credit from the bank which is supplying the finances. This letter can be written in such a way as to guarantee you payment from the bank through the developer or owner once the project successfully completed. Along with this, one contractor I know has gone to the bank of the owner or developer and asked the bank to also sign the construction contract, thus acknowledging their responsibility as the financial backer of the project and ensure that he, the contractor, receives payment upon completion.

A second method is to ask the owner/developer to supply a payment bond which you will pay for. As a contractor, you are often asked to supply this kind of bond which guarantees you will pay for the costs of the project or, if you default, the bonding company will be responsible to pay off all the outstanding bills and collect from you. This bond works in the same way. This time, however, the bonding company will stand for the owner or developer and guarantee if he doesn't pay you, they will. The bonding company will then research the reputation, credibility and financial strength of the owner or developer. If the bonding company, after they have done their research, does not want to take the risk and bond him, you had better think twice about taking a risk they will not. However, if they are willing to bond him, he is probably a good risk and will most likely pay his bills. At this point you may or may not want to pay the price of the payment bond.

Whatever you do in pre-qualifying owners and developers before you work for them, do it well and spend considerable time, making sure they are good and trustworthy people. Once you start putting time and money into their project, you are committed, and if they don't pay, all the hassle, energy and legal fees to collect your money (if you ever do) will suck every penny of profit and then some, out of the project.

Quality

Every contractor I have known wants to be known for his quality. However, quality is more than just a word, it is the reality of a job well done, and it is this reality which often escapes the people who are actually out in the field doing the work. What can be done to make quality not just a word, but reality in your firm?

Here are some ideas we have found being used by contractors throughout the country. First, start a program which emphasizes and continually brings to the attention of your people the desire of your company to have quality as a way of business life. The program may include a constant barrage of catch phrases which are posted on every piece of correspondence they get, and on every bulletin board and wall they look at. One of my favorite catch phrases in this price conscious business is "the quality is remembered long after the price is forgotten." These types of phrases can become the mottoes of your company.

Second, establish concrete ways to measure the quality of your work. One of the best is to pay an outside committee of judges to look over your projects when they are finished. It may cost you a few hundred dollars, but the understanding it will be done and the reputation it can give you with owners will possibly repay this investment many times over.

These judges need to rate every project in certain important areas on a scale of one to ten. The results of their scores will be given to you so you can carry out the third part of the program: Rewards!

Most programs fall by the wayside unless they have two important ingredients. The ability to measure someone's performance, and rewards for people based on that measurement. Some forms of reward, cash bonus, trips, "that-a-boys" in front of their peers, etc., must be a part of this program if it is to succeed.

Finally, know the program must continually be worked if it is to succeed. Recently, I was in a branch offices of one of the largest contractors in this country. I saw a sign with a quality motto on it which showed they were interested in quality. I asked the man I was with if they had a good program for quality. He said they had a great one. They had invested thousands in putting it together and getting it off the ground. However, he said it had fallen short of what it could do because no one was working it on a continuing basis, day in and day out.

Quality can and should be more than just a word, but if you want to it to be, take heed of some of these ideas.

Reinforcing People

One of the most important things we can do for the people around us is to reinforce and encourage them. Everyone needs this kind of help in our lives from time to time. However, the great majority of us are not disciplined in the practice of providing encouragement or "warm fuzzies" to our employees, friends, families, suppliers, clients, etc. Some of us get so wrapped up in our own world and the problems we are facing, we forget about or don't bother to think about the people who have made us successful. It seems for a very large majority, the only time we remember them is during the Christmas holiday, and only then because there is something that slaps us in the face and forces us to remember.

But here is an idea ALL companies could use, not just companies within the construction business. It is an idea revolving around compliments, satisfaction and discipline. It is an idea which, if followed very closely, will bring satisfaction to many people, including you.

The first thing you do is buy a calendar. Make sure it is one which provides enough space under each day. On the left side (which would normally be Sunday) write the word "Client" in the first week, "Employee in the second week, "Family" in the third week and "Supplier" in the fourth week. In each day of the week, draw a horizontal line through the middle of the block for that day.

Now here's what to do: Each weekday, go out of your way to compliment someone about something. It does not have to be a major item, but do not make it too simple. Once you compliment that person, write their name on a space on the calendar. Then continue this for each and every weekday until the calendar is filled .

Sometime during the week, send a "Thank You" note, letter of appreciation or send a small gift to one individual who falls under the category you have written on the left side, i.e. family, employee, client, or supplier. Again, if you send a gift, it does not have to be a large or expensive gift, but just something that says, "I appreciate what you do for me!"

Many of you may think this is a troublesome idea. But the important thing to keep in mind is that this will not only let your employees, clients, etc. know you appreciate them, but will also make you feel better. The key is to discipline yourself toward making this work. Your degree of self-satisfaction will be unlimited.

Rewarding People

In a previous section this book, "Bonus Systems: The Good, The Bad, and The Ugly," I talked of the dangers of bonus systems. I gave two major premises about them. First, any bonus system should not be able to be figured out because you will figure it your way, and employees will figure it their way and you will never have the same figures. Second, it should never be tied into measurable profit, because then you have taken on partners who will be after you about the way you do things and what you pay yourself so they can enjoy a part of maximum profit. However, I do strongly believe in sharing the wealth and rewarding people as a motivation for them to be on the team.

In this book I have already mentioned several ways of rewarding people for ideas and quality. But in this idea I want to talk about what we have found to be the best and most motivating reward of all, recognition. In the chapter, "Recognition For A Job Well Done," I outlined a program one contractor uses to maximize his recognition. We took a survey in a room full of employees on contractors at one of our brainstorming meetings. We asked them to rate what motivated them the most. As choices were gave them money, gifts or trips, or recognition for a job well done. It was interesting to note that money came in a strong last of the three by all but two people. The clear winner in 70% of the responses was "recognition for a job well done."

Then many of them spoke up and stated the most exciting thing to them was to have someone acknowledge their contribution to the company. They loved it when someone would point out to them that what they did and who they are were vital to the ongoing success of the company.

Here are some ideas to help you make sure such rewards are a part of your company. First, make a commitment you will personally acknowledge the contribution of 20% of your employees every week. That way, no employee will go more than six weeks without an "atta-boy." This can be done in several ways, but we were told one of the best was a personal handwritten note to an employee. One employee shared how he has kept every one of those notes writing by his boss and sometime, when the going gets tough, he pulls some of them out and reads them again.

Second, at meetings or company functions, be sure you find a unique way to acknowledge everyone's contribution to the company. What I am about to mention may sound corny to some of you but it gives you one idea of the kind of thing you could do to reward everybody. At a Christmas party for everyone in your company, pass out a rose to every employee in your company. At a given time during the program which you acknowledge the employee's part in the business, introduce each one and tell a little about his or her role in the company. As you do, have them come forward to put the rose in a vase and receive a gift. When everyone has put their rose in the vase, point out it is the small but important roles each person plays which contributes to the entire company becoming a bouquet which brings pleasure and success to those in and around it.

Someone once said the greatest reward is a job well done. I agree! But this reward is often short lived and shallow if the reality and recognition of it is not shared with other people.

The Scoopline

Every company has a grapevine upon which all the news, whether it is true or not (and usually not) goes around to each person. One of the deadliest things going around a company is gossip. People do strange things, and morale can take a dramatic dive when people hear you are almost out of work and they are going to get laid off. Or the company is unable to pay the bills and will be going into bankruptcy at any moment, etc., etc.

Often you do not even hear about this negative gossip or when you do, much of the damage has already been done. The key to keeping this gossip in check is to be very transparent, so everyone knows what is going on and will not believe the gossip because they believe if it was true they would have heard it from the proper sources.

One contractor has attempted to accomplish this through the use of what he calls "The Scoopline." He uses a short newsletter which goes to everyone with their paycheck. This newsletter can be used to build esprit de corps among the employees as well as squelch any possible or existing gossip with the truth.

He uses it to inform people of such personal things as who just had a baby, got engaged or married. He uses it to inform people of new contracts signed, new equipment purchased and new people hired, as well as positive events which are happening with the company. He also uses it to confront any negative gossip head on with the truth. His people have come to rely on it and trust it for their final source of what is really going on in the company. It can also discourage those who would start rumors and gossip because they know their prospective audience is going to weigh what they say against what they read in "The Scoopline."

There are two key items to making this type of program work for your company. First, it needs to be accurate. Even if there are some tough things you must reveal in such publication, you must be accurate. If it cannot be trusted, then it will become, in the minds of the readers, no more trustworthy than the grapevine, and its purpose will be defeated. Secondly, it must be entertaining and fun to read. If it is nothing but propaganda and information put in a matter-of-fact way, it will soon be discarded no matter how truthful the content is.

Remember, you do not want the truth about your company to be left up to the grapevine.

Spray Painting The Job

I ran into a landscape contractor who has a unique way of both selling a job and/or laying the job out for his crew after he receives it. I don't know if this method could be used in every case to sell a job, but it should be able to be used in every case to lay out all or part of a job.

When he went on a sales call, he would carry several cans of difference colors of spray paint. If the customer would permit him to, he would spray paint on the ground in the shapes of beds, the possible location of plants, walkways, fountains, etc. This gave the customer an actual feel for the size of some of these things and a better feel for the finished product than they would get from a scaled drawing.

Sometimes he would follow by getting their preliminary approval of the rough design on the ground, draw up a set of plans and then give the an estimate. If they bought it as sprayed out, the job was then already laid out for the crew. Any changes would be raked over and re-sprayed prior to the crew coming to do the work.

He reported on some occasions he was able to give the customer an estimate after spraying, and could then get their signature on a contract just from what they saw painted on the ground. He stated, however, this usually only happened on the small jobs that were not worth his drawing up a set of plans just for a sales presentation or for construction.

This is another example of being innovative. While this idea may not work for other forms on construction, it is an example of finding a way to be unique in our approach to business.

Survival

From the very earliest seminars I conducted, I have emphasized this is a very survivable business. It is not a business for the naive who think a few successful jobs and a few successful years mean they are destined to always be successful. There is only on thing for sure in this business: there will come a job or a time when you will need to apply all your talent, knowledge and skill just to survive.

With this thought in mind, I heard an important idea for the philosophy of management. A friend of mine has this philosophy: He believes it is the responsibility of management and ownership to make a construction company survive.

This means every decision management makes is made with this goal in mind. When it comes to buying new equipment or hiring new people, management understands its role is to make sure the company survives. When it comes to signing contracts or venturing into new divisions or work disciplines, management thinks it through with the idea that their decision is made based on survival. If it decides to build offices or open offices in distant locations, or enter into joint venture agreements with other contractors, the management must understand they are responsible to the company for its survival.

I believe if management would see itself more in this light, there would be more contractors who survive the good and the bad times.

Supervision Guidelines

One of the major problems in this business with both production and quality is the lack of supervision. Many times this occurs because you don't have qualified people out in the field to supervise. Other times it is a problem you inflict upon yourself because you spread your supervision on a project too thin.

How many people should one supervisor be responsible to manage? The usual figure we have gathered from others in the industry is one foreman/supervisor should not have to look after more than five other people at one time. That is the ideal ratio these companies strive for in the field (without exception).

We all recognize we are not always able to have the ideal. However, I feel if we don't have an understanding of the ideal and goals of reaching them, we will not achieve them most of the time.

The Level Of Success

I now talk of an old concept using a new term which I think may communicate this idea in a good way. I mention it here to remind you of an important old concept with a new term. The new term is "Finding One's Level of Success."

I believe everyone has a certain level of business and management we are successful performing. To do less than this level will mean we are not utilizing our abilities and opportunities to their fullest potential. This condition is not as dangerous as it is sad. But the other possible condition is not only sad, it is also dangerous. It occurs when a person grows their business beyond their ability to manage and finance the company. One of the things I find over and over again as I travel the country as a consultant, is many contractors fail because they get beyond their level of success.

I want to point out some of the reasons why or how contractors put themselves in a position to go beyond their "level of success." First, they find a good estimating system which becomes very successful in the marketplace. Because they are able to bid and get more work, they think they should keep up the pace or else they will not be living up to their ability. There are many jobs you may bid or want to bid, but in reality should be passed up because they will take you beyond your "level of success."

Second, many contractors are adventuresome people. They like to take on and try new things. I don't feel it is wrong for a contractor to venture into new territory as long as he limits these ventures, and makes them an exception rather than a rule. Many of you are successful at certain types of work, but those things do not hold great challenges for you because you have mastered them. Don't ever give those things up for the sake of a new venture, no matter how challenging and exciting that new venture may seem.

Third, and most importantly, they may have never been brutally honest with themselves about what their "level of success" actually is. Too often contractors do not pre-plan their business, or even have any direction or volume projections for years in business. They have no strategic plan for their company nor have they set goals for exactly what they intend to do as a company. This causes them to take the approach which says "Whatever comes and whatever happens, will happen." This kind of attitude sets them up to go beyond their "level of success."

I cannot emphasize this enough! Each of you needs to take control of your company's destiny and make it what it must be for you to be successful. Do you know what your "level of success" is, and are you working to make your company successful within that level?

Time Card Elimination

One contractor we know went on a campaign to reduce the volumes of paperwork which streamed through his office and eventually took up rows of filing cabinets for storage. One of the areas he went after was the time card systems we were using. The plan he came up with not only reduces the amount of paperwork but seems to make it easier for the field to report their time. It also gives a contractor who has crews working many miles from the office the ability to report their time on a daily basis.

Here is what he did. He added a telephone line into the office which was dedicated to payroll only. On that line he installed a highly efficient answering machine. His supervisors or foremen call this number any time after the workday ends (at their convenience but before 8:00 AM the next office workday). They record the workers' names and work functions they performed, and their hours from the workday on the answering machine. The person in charge of payroll listens to the tape, and enters into the computer all of the information which was given over the phone. The tape is kept for a period of time to make sure there are no hang-ups and the computer information (as always) is backed up once it is entered.

If you use this form of time recording, I would also recommend some form of release be put on the back of payroll checks so when an employee endorses it they will have acknowledge their acceptance of the hours for which they have been paid. Your attorney would know the best method within the particular laws of your state.

Another contractor uses a similar system, except he does not use an answering machine since his people all work in the vicinity of his office. He provides a small portable tape recorder to each of his foreman/supervisors. Twice a day, usually at lunch and end of day, they record the workers' names, functions performed and time. They can also use the tape to inform the office of anything which may be of importance or interest, and use it to dictate letters, change order requests or anything else the office can dispatch for them.

Time - Spending It Wisely

How should upper management spend their time so it will be most productive? Certainly, we are often the highest paid people, so we should also be the most productive. Yet, I find in many construction companies, upper management can get lost in all the things they feel they need to do and produce very little which brings back income to the company.

I have always had a gut perspective on this matter. Every time I walk into my office I ask myself, "What am I doing today which will generate $_____ for the company?" This kind of hard question first thing in the morning always seems to put what I decided to do with my time in better perspective.

I have discovered people in management who do not balance their time between several issues in a day are not as income producing as those who do. We can sometimes get so focused on a single project we let other things slip into the next day, then the next, etc. Here is a formula which will help balance your time, especially if you are a smaller contractor.

Spend one-third of your time *Finding*. Finding new work to bid on and to contract.

Spend one-third of your time *Minding*. Minding the work you have under contract and making sure it is getting done properly.

Spend one-third of your time *Grinding*. Grinding out all the little things which need to be done, most of which none of us like to do because it is the repetitive type of work which tends to be boring.

The best way to discipline yourself to do this is to pre-plan your days as far in advance as you can. Then look at those plans and see if you are spending your time as we just described. However, to plan it is not enough, because our days don't always work out like we planned them. So, as your days are completed, record how you actually spent your time and see if you are still following the formula.

Travel Pay

One of the most difficult items to deal with in this business can be that of paying people to travel to and from a job. If you have read the rest of this book, you know I am dead set against having people come to your shop and be assigned a job to go to, and be paid to stand around waiting for the assignment, and then be paid to travel to it. As much as is possible, I always recommend people arrange their own transportation or be told to carpool from the shop in your trucks, or their peoples' cars, to the jobs.

However, I know for some of you this is an impossibility right now. You have either been doing it another way and it would be hard to just change overnight, or, you have such small jobs you cannot know who would be going where until that very morning when you are doing the schedule. Or, you are in a state where the labor laws insist you must pay travel time. Here is an idea about what one contractor is doing, one you might want to consider because it can help bridge the gap between your present situation and where you would hope to get to, and meet the requirements by law.

He is paying a reduced rate for travel time. The rate can be just above minimum wage so you are within the law. This can be justified because you realize they deserve something for their time to get to the job, but they should also realize they are not actually working on the job, thus the rate of pay should be reduced accordingly.

Vehicle Maintenance and Responsibility

One of the biggest problems we continually hear about from companies is about vehicle maintenance, upkeep and the extensive damage which seems inevitable. It seems a company's vehicles are always "getting destroyed," yet none of the employees know how it happens. These "accidents" can quickly drain your checkbook. Although you use a method for figuring the cost of maintenance for your vehicles/equipment (or at least you should be), how can you go about recovering the cost of the dents, scratches, broken parts, etc.?

The solution is quite simple. In order to keep the vehicles and equipment in the best shape possible, make the users responsible for them. You may be saying to yourself, "Well, I already do that." But you need to take it one step further.

Begin by developing a card, placard or notebook for each vehicle and large piece of equipment. On this card, write a description of the vehicle to include make, model, year, etc. Below this will be a line item for each of the following: oil, fuel, transmission fluid, brake fluid, windshield wipers/washer fluid, headlights/tail lights/brake lights, emergency flashers and turn signals. Below these items or on the back side, make a two or three line space of the following items: exterior condition, interior condition, dents or scratches, tires, comments. Below these items, make a matrix for each day of the month (or a calendar). Then place the card in a notebook or clipboard inside the vehicle.

Instruct your vehicle/equipment operators that at the beginning of each day they are to take this card and make a complete inspection of the vehicle/equipment, both inside and out. After the inspection, they are to place their initials in the appropriate block for the day of the month they are using it. If there is any major damage spotted which has not been previously noted, the operator is to notify you immediately.

Assign one person (a supervisor) the responsibility of making random checks of each of these cards, either daily or weekly. The individual will be checking to insure the card was initialed for the appropriate day, and any damage/problems were noted accordingly. Anything not annotated, should be brought to the attention of the vehicle/equipment operator. If there is any damage to the vehicle which was not written down or reported, the last person to initial/sign for the vehicle will be held responsible for the cost of repairs.

Although this may seem troublesome at first, you will be amazed at how well your employees take care of vehicles and equipment when they know if they break it, they will be responsible for it. This process can save you literally hundreds of dollars per year in vehicle/equipment repair and replacement. A word of caution concerning this suggestion: prior to trying to enforce a "you break it, you buy it" type of rule, consult with your attorney regarding the legality of such a policy. One of the keys to making a program such as this successful is to try to have your employees respect your vehicles and equipment. But you do not want to get them to the point where they will not touch any equipment for fear of damaging it.

Sample Vehicle Log/Discrepancy Sheet

Year: Make: Model:

Plate Number   Month:

1. OIL: check oil level, check for water, cleanliness
2. TRANSMISSION FLUID: check level, check for water
3. BRAKE FLUID: same as #2
4. WINDSHIELD WIPERS/FLUID: condition of wipers, fluid level
5. HEADLIGHTS: both low and high beams operational
6. TURN SIGNALS: operational
7. BRAKE LIGHTS: operational
8. EMERGENCY FLASHERS: operational
9. GAUGES: fuel, oil pressure, battery, water temp., etc.

A. Interior Condition:_____

B. Exterior Condition:_____

C. Dents/Scratches - Location:_____

NOTES

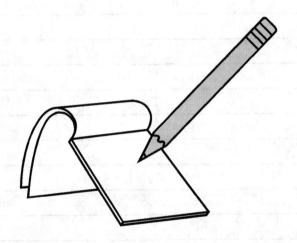

Troubleshooting Your Contracting Business to <u>*Cause*</u> *Success*

Troubleshooting Your Contracting Business to <u>*Cause*</u> *Success*